What Readers Are Saying About *Edge of Yesterday*

"Finally, a story starring a girl who's seriously into science! *Edge of Yesterday*'s Charley thinks about more than just clothes and boys. A must-read, especially for girls who feel their interests don't match those of their peers."

—Valerie Wang, 14

"*Edge of Yesterday* is an outstanding book with so many amazing characters! I would've stayed up reading it all night if I had the chance."

—Gigi W., 12

"At a time when girls are underrepresented in the STEM field, finding a book that features a girl with enthusiasm and energy for engineering is really encouraging. The perfect amount of drama, science, and quirky and relatable characters—you'll feel an instant connection to Charley."

—Rita Zhang, 15

"Charley is a perfect example of a smart girl who's not perfect, which makes her more relatable. Charley's passion for time travel and Leonardo da Vinci inspires me to be more innovative. I'm excited for the sequel."

—Mayo Olojo, 16

"Charley's love for engineering is inspiring to girls facing obstacles while trying to follow their dreams. Realistic characters make it a fun read and leave you wanting more."

—Katheryn Wang, 16

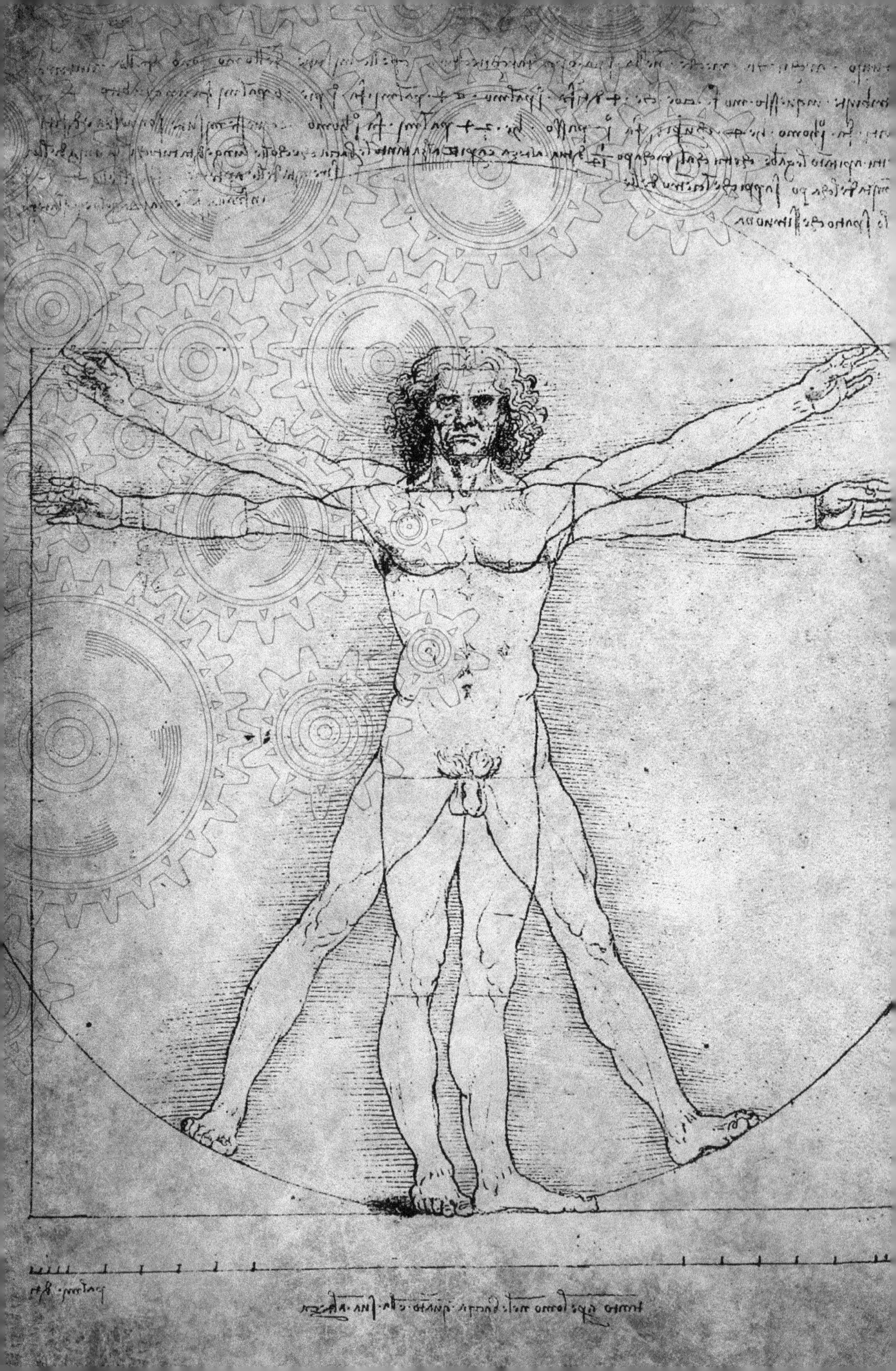

Edge of Yesterday

EDGE

OF

YESTERDAY

~ a novel ~

BY

Robin Stevens Payes

Cover design by Melissa Brandstatter
Interior design by Lisa Vega

Author photo by Judy Gee

Photos of Leonardo da Vinci documents on pages 8 and 13, courtesy of
Luc Viatour/Creative Commons/Wikimedia
Other images courtesy of Creative Commons/Wikimedia and Robin Stevens Payes

Library of Congress Control Number: 2017934042
ISBN: 978-1-937650-83-4

SMALL
BATCH
BOOKS

493 SOUTH PLEASANT STREET
AMHERST, MASSACHUSETTS 01002
413.230.3943

SMALLBATCHBOOKS.COM

Dedicated with love to Ben, Dana, and Ari,
who inspired me on this journey to the Edge of Yesterday.
And to Rick, who gave me the courage and support to believe I could
write fiction, I am deeply grateful.

"Time stays long enough for those who use it."

—Leonardo da Vinci

Contents

I.

OF NOTEBOOKS AND FLYING MACHINES

The sun is bright as we scramble outside the Smithsonian Metro station. I lower my baseball cap against the glare. The National Mall—the nation's front yard, as they call it—is jam packed, as always, when the weather's as perfect as today. Joggers dot the gravel path, while people out to enjoy the day pass more than a few Frisbees and footballs overhead. Big brown spots reveal where too many visitors have trampled the grass over a hot, dry tourist season.

Now that it's after Labor Day, hordes of students and interns from all over join the throng of tourists enjoying these last warm days. Teens pretend not to know the grownups trailing behind them pointing out historical landmarks. Even though Washington is not a majorly polluting city, tour buses kick up fumes. My science fair partner, Billy Vincenzo, and I are drinking it all in. Oh, and did I mention? We brought my dad, the science nut, along for good measure.

Allow me to introduce myself: Charley Morton, at your service. *Servizio,* they'd say in Italian, which I'm trying to learn. I'm just your average thirteen-year-old girl with wild hair, freckles, and braces (with purple and gray bands—school colors).

There's so much I want to explore in this world. I love the logic of science, math. The beauty and precision of language. How did that evolve, anyway? I want to discover the music of the stars and figure out how to extend human life spans in good health. Through art and anatomy, I'd like to understand the mechanics of the human body as a player moves down the soccer field. Not just to see how a body in motion works, but to improve on speed, economy of motion, and accuracy of the kick. And through forensics, I want to learn why every human fingerprint is different.

I want to do it all!

That's why Leonardo da Vinci, the ultimate Renaissance man, is my personal hero. He's also a big part of why we—Billy, Dad, and I—are visiting the Smithsonian today.

"Hey, Charley," Billy says, "take a picture for me, would you?" He uses his antique flip phone to point to the Museum of the American Indian looming ahead of us. "The new game I'm building features a virtual world among cliff-dwelling Indians. You know, the Anasazi? I'm gonna make scenes with these prehistoric people living in caves on high cliffs in the Western Plains—scaling the steep canyon walls to get home, hunting, fighting. But seeing the museum design here, in person, I'm like, how could they possibly have carved a village into those rocks?"

"You are such a fossil!" I tease, pulling out my cell phone and waving him to move into the picture. "Why don't you get a smart phone and join the twenty-first century?" But I know that Billy's parents are afraid he'd spend all his time playing games. Or designing them. Which he might.

Most people think Billy's just a nerd, but I think his brainiac-ness is cool because he's smart like me, only way more focused. Billy definitely qualifies as Da Vinci Middle School's geekiest eighth-grader. Some people think I'm the nerd, but Billy is more of a gearhead than I'll ever be, because he's got a one-track mind for the tech stuff, whereas I'm determined to study, well, everything. How else to strive to become a modern-day Leonardo da Vinci?

Exploring the National Mall, which stretches from the Lincoln Memorial to the Washington Monument, past the Smithsonian museums, and on to the Capitol building, makes me happy because it's like walking in the footprints of history. I imagine the gravel path running parallel to all the museums as it might have been two hundred years ago: horses and buggies kicking up dust; the Capitol building in cinders after the War of 1812; men sporting top hats and canes, and corseted women carrying parasols to shade them from the hot sun.

But it was probably nothing like that. What I'd really like is to see these things for myself. If I could, I would interview all the brilliant minds throughout history who, back in the day, radically helped shape the future. In my book, it starts with da Vinci, of course. Then Newton. Copernicus. Darwin. Elizabeth Blackwell, first female doctor in the U.S. Einstein. Marie Curie. And true to this very place in the nation's history, Martin Luther King.

Meanwhile, here I am in cutoffs, tie-dyed T-shirt, and sneaks, my wiry hair momentarily tamed into a ponytail under my cap. I've got my tablet—as always—in my backpack and cell phone in my pocket at the ready so I can take pictures for my blog.

This day belongs to my favorite guys—Dad, Billy, and Leonardo. I don't come inside the Beltway often; for one thing, I'm not great in crowds. I can get overwhelmed, and sometimes I get so turned around that I can't figure out which way is up. My parents had me tested after I got lost one too many times on the Metro, and the verdict is I'm "directionally challenged." Not great for someone who dreams of making distant travels—through time, no less.

I hear the beautiful ping of a Snapchat and pull out my phone. Dad gives me that frown.

"Charley, you know—no texting once we get inside!"

"Duh!" Does Dad think I have no manners at all, I wonder? But I do have to check, because there's the matter of Beth, my once-upon-a-time best friend who has suddenly gone rogue.

Sure enough, Beth's Snapchatted me a selfie looking all Bella Swan—hair dyed blue-black and freshly straightened, eyes super made up, and waving newly manicured blue nails with little orange dots on them at the camera.

Billy walks over and looks at my phone. "Hey, what's up with Beth—she dressed up for a costume party or something?"

I frown. "Who cares?" I try to throw the question away.

But I do care.

"Say, you two," Dad, who's been continuing on toward the Air and Space Museum, turns back, seeing we're halted. "Let's not get sidetracked, guys. Something wrong here?"

I stop to think briefly about explaining. But Dad and Billy are both clueless when it comes to the intricacies of teenage girls' minds.

"Never mind." I shrug off Bethy's weird behavior and start down the Mall in a sprint, Dad and Billy following in my tracks and straining to keep up.

Dad, panting, stops us right at the foot of the steps to the Air and Space Museum. He really oughta start working out more.

"Hungry, guys?"

I'm always hungry! Looking for food here makes me feel a little like a modern-day urban forager because of all the colors and scents from the food trucks that line the streets around the Mall—Lebanese, Thai, Korean, Cajun, and deli, for starters. We all opt for kebabs at the Lebanese truck.

Then, the kid stuff: ice cream, cotton candy, cookies. Rocket Pops are my favorite cool treat, so I'm glad I brought my babysitting money along to indulge (the parents strictly limit my sugar intake, so if I want something special, I have to pay out of my own savings—like a little candy now and then is gonna kill me!). I love sucking the juice out of those red-white-and-blue rockets.

You may think I'm weird, but I collect the sticks, and my friends and I make up fortunes to write on them . . . sort of a fun doodle project. "I'm gonna write one that says, 'You are going on a long journey,'" I quip, reflecting on my hopeful plans for somewhere warm during winter break. I stick it in my pocket to add to the collection in my Girl Cave at home.

In today's heat, I have to chomp down quick on the ice before the sticky stuff drips semi-permanent rivulets of cherry syrup down my arms. I start biting, braces and all, and—brain freeze! I pop off my Washington Nationals baseball cap for a quick thaw.

Billy, orange Push-Up pop momentarily forgotten and dripping onto his shoe, is squatting on his haunches, studying the refrigeration coil system on the ice cream vendor's ancient pushcart.

Dad looks from Billy to me with a quizzical expression. (Quizzical, isn't that a good word? It means mystified and slightly amused—like a combo of puzzled and comical.) Meanwhile he munches on his popcorn—one annoying kernel at a time. Between the two of them!

"Can't you hurry, guys?" At this rate it feels like we'll never get to the Leonardo da Vinci Codex exhibit, so I start skipping up the steps to Air and Space. Dad has to take big steps to keep up. And Billy . . . well, suffice it to say we keep almost losing him to myriad distractions before we even get inside the doors of the museum!

Once we do make it inside, I hurry past the World War II exhibit and the moon lander. Those I can see any time. Leonardo's Codex is on display in the same room as the Wright Brothers' first flyer.

I hear myself babbling in my excitement. "Everyone calls it a notebook, but a codex was a real thing back in the day," I explain to Billy.

"Definition: an ancient bound book made up of separate pages. In fact, some of these pages consisting of sketches and descriptions were probably assembled into books after Leonardo's death."

"Um-hmm," Billy murmurs. He's peering up at the wings of an early glider. Not listening.

"A codex would be like journaling today. Or blogging. You know, like Ms. Schreiber's asked us to do for the science fair project?"

I slyly check my Instagram—another way Leonardo da Vinci might record his observations today—and there's a sullen selfie of boy-crazy Beth in her room at home with a stuffed lion in her lap and holding up a piece of paper: GROUNDED. It's all about her sneaking out with Da Vinci Middle School's chief stud and star athlete, Lex, during what was supposed to be a sleepover at my house. So that's that. I think briefly about it—how once upon a time we would've all been science fair partners and she would be here, too. But then I remind myself: *That* Beth is gone.

"Hey, Charley," Billy calls, peeling off to get a better look at the Wright Brothers exhibition. "I just want to see this one thing before we

get to Leo—the Wright Military Flyer. It's the first military plane ever; it first flew a hundred years ago. Hold up!"

"Only a hundred years? That's nothing!" I call behind me.

Billy seems stuck on the Wright Brothers. "What about the science behind manned flight, Charley? You know, for the science fair? After all, Leonardo may have dreamed it, but Orville and Wilbur actually made it happen."

"Yeah. Maybe," I reply without breaking my stride.

"Slow down, Charley," Dad says, tugging on my arm to stop. "Let's not lose Billy."

"Can't wait!" I say, running ahead, for once not caring to keep track of the world's biggest space cadet.

Almost there. I'm aware there's a crowd, but I don't feel it. I slip between a couple of "suits" as Dad would call them—those official Washington types: diplomats, lobbyists, and senators—and squeeze past a pod of pale tourists who are staring into their guidebooks, oblivious to the genius before their very eyes.

As soon as I find myself in front of the Master's work, I am speechless.

Up front, I take out my phone to start snapping pics. Immediately, an eagle-eyed guard shuts me down: No photos allowed, he says, not even when I explain that I am Leonardo's biggest fan and that the pictures are strictly for educational purposes.

I glance from the Leonardo notebook app on my tablet to the REAL THING; the actual Codex is light years more awesome when I am standing barely an arm's length away from it. And the whole thing is only about the size of a cheap paperback.

But the contents of this ancient paperback are here before my eyes! Sketches and diagrams in the *Maestro*'s own hand! Mirror writing—that was the real da Vinci code—writing from right to left instead of the normal way. Presumably, this was to confuse people who might steal his ideas. But I think it might've been more because he was a lefty and didn't want to smudge the page as he wrote and sketched. I have the same problem.

And truly, without being able to understand the Italian dialect of the time or being able to hold up the finely written script to a mirror, it would be impossible to decrypt what Master Leonardo wrote and doodled just by looking at it. So the museum curators have been thoughtful enough to translate Leonardo's cryptic handwriting into English as part of the exhibit.

Somebody's poking my shoulder, and here's Billy, panting. "Charley, couldn't you even stop for one minute to see what I'm interested in? I mean, really. Why I even bothered to come down here with you in the first place is beyond me."

"Science fair, Brainiac," I reply, giving him the eye. "Did you forget what we are here for?"

"There's other stuff that's science fair–worthy, *Brainiac*," he replies. "Weather balloons, for example . . ."

"Well, if you don't like the idea of recreating one of the experiments of the ultimate Renaissance genius, maybe I'll just find myself another science fair partner. . . ."

"Why are you being so stubborn, Charley?" Billy says, his voice getting louder. "All I'm saying is—"

"I know what you're saying, Billy, and you're missing the point, which is right in front of our noses."

"What point? That Leonardo da Vinci thought he could fly four hundred years before the Wright Brothers did but didn't have the technology?"

"Well, for starters, yes," I say. "If you'd only pay attention . . ." Suddenly, I notice two of the "suits" staring over at us with a disdaining look.

Before I can continue, Dad manages to squeeze his way within earshot. His eyebrows are pulled together angrily. "Charley, you see those guards over there? They are ready to eject you from this room. So I suggest you two stop bickering and look quick. There are fifty people behind you now waiting patiently for a close-up view."

"I'm gonna go check out the military plane again anyway," Billy whispers, drifting back toward the Wright Brothers exhibit. "'Cause I think it'd make a cool project for Schreiber's class."

"Hmmph. Once-in-a-lifetime opportunity. And you're about to miss it, Vincenzo," I mutter as he walks away. I take out my tablet to tap in a few notes.

I turn back to the book in the glass case and draw in a breath. Its size makes it hard to study unless you're standing virtually on top of the case. The pages are yellowed and spotted, as you might imagine of something that old. And some of it shows the ghost of older writing erased or written over—a palimpsest. (This is what they call a manuscript or piece of writing material with the original writing erased to make room for later writing, though traces remain.) Because paper and parchment cost so much, and Leonardo was not wealthy, he would've reused the paper. I've read that you can even see his grocery list on some pages. Imagine peering at what was supposed to be notes on flight to find out what he was buying for supper!

Not here, though. These pages show how carefully Leonardo observed the world around him and how meticulously he would capture such things as the array and composition of feathers on the wing of a bird. Or the swirl of currents in the air, that no one could really even see.

"To think he was given the time and freedom to just be curious, to explore, investigate, and follow his passion!" I say wistfully, immediately resenting how much of my time is taken up with standardized tests, or memorizing historical facts that anyone could just Google. In fact, I've read Leonardo wasn't even allowed to go to school.

"I wish I'd grown up in Leonardo's time!"

Dad, having heard me on this subject many times before, stops me. "Now, Charley, we've been over this. Italy in Leonardo's day was barely out of the Middle Ages! Leonardo had the 'freedom,' as you call it, because there was no such thing as public schooling back then. In fact, your Leonardo had to teach himself."

"Wonder how many languages Leonardo knew?" I screw up my eyes trying to decipher Leo's mirror writing for myself.

"Whatever form of Italian was popularly spoken in the fifteenth century," Dad replies. "Plus, the highly educated would also have to know Greek, Hebrew, Latin—and your Leonardo would have read some of the classics."

I consider this as I inspect the drawings of birds and bird wings. This could be the sketchpad of someone whose main preoccupation was drawing, which is the one factoid most people know about the artist who created the *Mona Lisa*.

"But Leonardo was clearly a scientist, too. Look at how detailed these drawings are! He even designed a helicopter—look, Billy! . . . Billy?" I turn and bump the arm of an art student sketching her version of Leonardo's sketches, and remember that Billy had already wandered away.

Dad scans the room looking for him. Luckily, Dad is tall and can see over most of the heads in the crowded exhibit. "Don't worry, I'm sure he'll be right back."

"If he doesn't get lost!" I respond, sighing. I turn back to read some of the English translation. As I try to edge in closer another sketch catches my eye—a pyramid-shaped scaffold with a semicircular track around it. "Wow, looks like you'd have to add 'engineer' to his list of accomplishments," I say.

"He was indeed a master across many disciplines," Dad affirms.

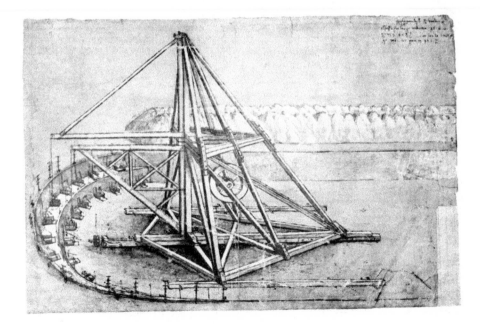

"The original brainiac!"

Dad chuckles. "In fact, his patrons were often angry at him for starting so many things he couldn't finish. Sound familiar, my little brainchild?"

"Dad, don't you ever wonder how he did it? I mean, surely there weren't enough hours in his lifetime to master everything he did!"

Dad just laughs. "Take the long view, Charley. You're only thirteen."

Walking backward to read the signs and notes along the wall, I sigh. "Sure would be nice to interview Leonardo for the science fair."

At that moment, Billy rushes back, excited. "Charley, wish you'd help me think this through. I'm thinking we could build a cool game: 'When Wars Took Flight.' Whaddaya think?"

"I think you're totes a gamer geek, is what I think."

I pivot back to study the notebook in the glass case. I try to fix the image of the Codex in my mind like a photograph, though I suspect even photographic memory would fail to capture every line and stroke of genius here. I hope just maybe I can become a little Leonardo-ish by staring at his mirror writing. Then I close my eyes to imagine spying into Leonardo's studio as he captures his world on the page. But there are elbows bumping into me, and I suddenly tune in to all the voices around me. My world, here, today, rushes back. I suddenly feel the crowd closing in on me. Too many people.

"I feel a little faint," I whisper.

Dad puts a protective arm around my shoulder and leads me away from the throng. "C'mon, Charley," he says. "Let's get you out of here."

"Uno momento, Papà," I say. I pull back to soak in Leonardo's *Vitruvian Man*. Today, it is seen as a symbol of timeless humanity—universal. *L'Uomo Universale.*

I sneak a single photo of the pyramid sketch and post it to Instagram with the caption: "science fair?" With such a big crowd, no guards seem to notice this time. "Sending to you for later," I text Billy with a copy of the photo. "Could be useful for our project."

Just then, Billy chimes in. He's back. "You know what would be useful, Charley? If you'd take a minute to look at the Wright Brothers before we go."

By the angry note in Billy's voice, I can see it would mean a lot to him. "Sure, Billy. Love to."

But Dad's got his arm around my shoulders, steering me full speed ahead toward the hallway. "Another time, guys. I need to get us home, and Billy to his mom's office."

"Umm, sorry, Billy. Let's come down again when it's not so crowded and spend time looking at what you want to see, 'kay?"

"Yeah, whatever."

"No, really, I promise we will!"

Billy, not completely convinced, trails reluctantly behind us.

"Imagine being the first to invent a flying machine," I resume, loud enough for Billy to hear. "Leonardo literally dreamed up his flyer almost five hundred years before your Wright Brothers, after making studies of birds' wings. Wonder what he'd make of how far we've come—to the moon, Mars, and now, out of the solar system."

I squirm out from under Dad's arm to walk backward, imagining that long voyage through space and time, even as we observe the history of flight unfolding in the exhibits around and above us. Billy puts his hands out to stop me just before I come crashing into him.

"Guys, do you think time travel is possible?"

Dad's left eyebrow goes up. "Hmm. I'm not sure I know how to answer that, *cara mia.*"

"I mean, hypothetically speaking?"

This piques Billy's excitement. He's all about the calculus behind the science. "Well, hypothetically, we don't know that it's *impossible.*"

"What if Leonardo actually created plans for a time machine? I mean, flying was only one of his obsessions. Because he was, like, obsessed with inventing the future."

Billy looks unimpressed. "You know who really invented the future, Charley? The filmmakers of *Back to the Future.* You know, October

15, 2015 . . . check out the future." He gestures around us.

"Yeah. Predicting 1985 to 2015. Big whoop." I glare at him. "We're talking 1452 to today! So wouldn't you expect, you know, he'd also try to invent a time machine to see this future?"

"Charley, the science and technology back then . . . I mean da Vinci was hardly a scientist!"

"He was all about observing the world around him. That's science, isn't it, Dad? That's what I think, anyway."

We're nearing the exit. I can't help taking a skip, and I almost bump into a stroller coming up from behind. Billy pulls me away just in time.

"Your da Vinci was definitely a man of science, Charley," Dad says. "But his methods, by today's standards, would be judged as a little loosey-goosey."

That cracks me up. "Loosey-goosey, loosey-goosey!" Outside, I begin waddling and honking like a goose, up and down the stairs outside Air and Space, pretending to tap innocent children on the head as I pass by. "Duck-duck-goose. Squawwk!" I laugh at my own joke, then wait for Dad's eyes to roll. I know I should act my age, but I figure I do enough of that with school and stuff. Sometimes a girl's just gotta let loose!

I spy two guys in Oregon Ducks jerseys passing a football. "Look, Dad. Loosey and Goosey!" I climb over the banister and slide down feet first, reaching the bottom of the steps at the same time as the football flying right at my head.

"Watch out, Charley—!"

"*Ooph!*" I hear a sound like air coming out of the fireplace bellows as I hit the ground.

II.
Veni, Vidi, What?

My hands and knees are all gravelly, but my first thought is for my tablet. I fish into my backpack and let out a sigh when I see the cover protected it.

"Charley!" Dad wraps an arm around my shoulders. "Let's get you home, champ." Billy hurries over with my cap, which must have gone flying when I did. Guess I got a little too loosey-goosey there. Making every effort to act more circumspect, I let Dad and Billy take the lead until we get underground.

On the subway, I rest the lump-free side of my forehead against the cool glass of the window and watch the blackness of the Metro tunnel become a lighted platform of passengers and then turn back into blackness.

I can't stop thinking; the Mall, the ball, the fall . . . but most of all, *l'Uomo Universale*, Leonardo. Before I know it, Billy interrupts my reverie.

"Guess this is my stop," he says. He's getting off at the Amtrak terminal, Union Station, to meet his mom. She works on Capitol Hill. You wouldn't know Billy's parents were any big deal to hear him talk about it, but I think she's chief of staff for a congressional committee or something. And that's a Pretty Big Deal in Washington.

"Thank you, Mr. Morton. See ya, Charley. Wanna get together at the library to talk about the science fair? Unless you have someone else in mind instead, that is."

"Might," I say. I mean, considering Billy's extreme resistance to all things da Vinci today, maybe I'd be better off with a different partner. But Billy *is* the smartest geek in the class, and we *do* make a good team.

"Nobody's gonna get you like I do, Charley," Billy says. "You know I like . . ." He starts to say something, but apparently thinks better of it.

"What. What do you like, Billy?"

"Not a what," he replies opaquely.

"Oh, I get it. You're a what. I'm a who," I joke, trying to lighten the mood.

Billy sighs. "The old joke is, 'Who is on first. What is on second.' But that's not what I'm talking about, Charley." He's got a strange, cat-like look in his eyes.

"Whatevs." In fact, I have no idea what he's driving at. "C'mon, Billy. We're gonna figure this thing out. You and me. Okay?"

"Yeah. Sure," he says. He throws his backpack over his shoulder, making his way to the opening subway door. "See ya, Brainiac."

"See ya, Brainiac." As the doors close, I see Billy standing on the platform staring in. The train begins to pull out of the station and I wave goodbye, then watch as his face turns from hopeful to disappointed.

What's that all about, I wonder?

I feel a slight jolt as the train picks up speed and rounds a bend that pushes me up against the window. Setting down my backpack, I adjust myself on the seat. We're lucky to have seats, I remind myself, seeing so many people standing in the center aisle, holding tight to poles and handrails. Wonder what Leonardo would think about modern subways, anyway? Bet he'd engineer a more reliable system than the D.C. Metrorail.

"Don't you think you were a little hard on Billy, Charley?"

I feel the rumble of oncoming trains rattling by on the opposite tracks. "Geez, give me a break, Dad. I just want him to be as excited about creating a project based on Leonardo's inventions as I am." We come out of the tunnel back into the light of day before the train squeals to a stop at the next station. The glare, plus the bump from that awkward football head-banging, are making my head hurt a little. I snuggle against Dad's shoulder and close my eyes, when it comes to me: an image of Leonardo sketching on parchment scraps at his drafting table.

He's attempting to capture the flight of some hawks circling in the sky outside his window—dashing off sketch after sketch of wing shapes and of birds soaring aloft. I follow his line of vision and spy the outstretched wings of those birds riding the thermals, soaring. I can feel my own presence in Leo's second-story artist's workshop.

Leonardo is no more aware of my observing him than the birds are aware that he is observing them. I peer dreamily over his shoulder as his designs emerge on the page.

I keep watching: Leo draws spirals and swirls—lines that would seem to illustrate the flow of wind. His sketches are precise. Despite noisy distractions from below—the laughing of children, a braying donkey, a woman from somewhere below us scolding someone—his focus never wavers. If he perceives a stranger in the room—a.k.a. *me*—he never lets on.

He takes the sketches and attaches them together with a leather strap, then tucks them under his belt. Apparently, the birds were just a momentary fixation. I notice the other sketches littered around the drafting table as Leonardo goes back to work on a detailed drawing— some invention or other—that he evidently set aside to capture the birds in flight.

With great steadiness of hand, Leo's working on the details of a pyramid-shaped scaffold design—the same design I saw finished not fifteen minutes ago at the museum! The pyramid shape is surrounded by what appears to be a semicircular track on which it might move into a tunnel entrance burrowed into a hillside—is it designed to dig deep into the earth?

After an intense focus that seems to last forever, he steps back, looking satisfied. Leonardo exhales in relief and puts down his pen, shaking out his hands to release the tension. *"Ecco."*

So this is the great Leonardo's studio. All around me are half-finished works—paintings, models for sculptures—and the tools of an artist scattered about: palette, brushes, chalk, oil paints and charcoal, a straight edge, a drawing compass—so golden and shiny with sunlight angling in through an unseen window, it catches my eye.

An easel is set in the corner with folio paper and notes tacked to one side of a charcoal drawing on canvas. The *Vitruvian Man*! I clap my hands over my mouth to keep from gasping at the nearness of it all.

But Leonardo doesn't seem to notice me. He brushes his cheek, leaving a black smudge, before wiping his hands with a rag hanging on the easel. After examining the fruit of his labors, he moves to the open window and sticks his head outside.

"Eh, ragazzo!"

"Si, maestro?" a voice calls back.

Am I dreaming? But this feels different—the smells, the sounds. There's a live drama happening in front of me. It's not a movie or a play or a dream; more like a real-life window into a distant past.

I move closer to the window—on tiptoe, so as not to disturb whatever pseudo-reality this is—to see what (or who) has captured Leonardo's attention. We are looking out over a main square in what I assume is Florence, Italy, at winding streets and alleyways leading from the square, people wandering on foot or steering peddlers' carts. Men riding horses that kick up dust, not to mention their poop, in the street.

Women in carts. The occasional over-the-top gold-painted carriage, no doubt carrying some noble person.

There's a platform beside a giant pyramid of stuff—what looks like gold, paintings, and silver—and that scaffold thingy. At the edge of the stage lies a guy caped in black, around my age I think. He appears to be sleeping.

Leonardo shouts, *"Eh, bene, Kairos, svegliati!* It is time to wake up. I need your help!"

The scene blurs. Dad's shaking my shoulder. "Hey, kiddo! Charley! It's time to wake up."

"Huh?" I manage. How'd Dad get here?

"We're here. This is our stop."

The train grinds to a halt, and I feel my body jerk against the seat. Hearing the screech of brakes as the train pulls into Takoma Park station brings me back to the here and now. Outside, the parking lot's agleam with shiny cars reflecting the late afternoon sunlight and Metro buses pulling in to the station.

The world I was in is not this world. But I swear it felt as real as anything I have ever experienced. I stand and stomp my feet to make sure I'm not dreaming. Something metal clatters to the ground. Must be my lucky Susan B. Anthony dollar. I get down on my hands and knees and turn on my cell phone's flashlight to search under the seat.

"Where's my Susie B?" I mumble, seeing only discarded *Washington Post Express* newspapers, leaves, and gum wrappers littered on the floor, until I touch not a coin but a golden compass like the one I saw in Leonardo's workshop, its dulled, golden patina—pockmarked and dusty— still bright enough that my flashlight picks up a reflection.

"Get up, sport! That floor's dirty!" Dad pulls me up by the hand, but not before I've grabbed the compass in my other hand. It's kind of like one of those old-fashioned skeleton keys. The weight of the metal feels

right—not too heavy, not too light—and somehow I know this is proof of where I have been, what I have seen.

I let myself be pulled along as Dad hurries out before the door slams on us. This compass, dirty though it is, appears to be rare, well worn, perhaps even an antique—not like the mass-produced model I use in geometry. There's something familiar about it.

"Attention. Doors closing," intones the recorded voice announcing that the train is about to depart. I slip the compass into a pocket of my backpack and zip it up tight. No one can see this memento from out of time—not yet.

"Good dream, sweetie?" Dad strokes my cheek lightly. "It's been a long day."

I have proof I wasn't dreaming. I don't dare tell Dad what I saw, but Billy's another story.

III.

On Top of Spaghetti

Coming back to Planet Takoma Park brings more of life's perplexities. (Okay, same as confusion—but it's a good word, don't you think?) Beth was following my Instagram posts, and now she's texting me about how Lex has agreed to be her science fair partner and they've got a very cool project idea.

Lex. Like I could care!

Tomorrow's a school day, and my head's just swimming. This new world of teen friends is way trickier than I would've expected. I cannot understand what's changed between Beth and me. Given the tension I'm feeling based on all the pings back and forth with my former BFF, I can almost understand why the parentals were so reluctant to give me a smart phone, or allow me on Instagram.

I wonder if this is something I could talk about with Mamma, but then I remember: She's busy getting ready to go on tour for three whole weeks with the National Symphony Orchestra—did I mention she's a concert violinist? It seems like they're touring all the time, but she insists she's here for us—Dad and me—when we need her.

And where's the NSO headed? To Florence, Italy, of all places. I sigh. Wish I could go!

I run upstairs as soon as we get home to make some notes in my journal.

Very excited about science project. Leonardo time machine? Explore possibility with Billy—time travel NOT impossible: What quantum physics principles would allow time to travel backward? And what's a golden compass got to do with it?

I'm too breathless with excitement to write any more than the basics right now. I take a wire brush to my unruly curls, hoping to maybe also untangle the fractals inside my brain, but my mind keeps skipping from the Leonardo exhibit to my so-called dream on the Metro and inexplicably finding the golden compass—which leads me back to Billy and his weird behavior and, from there, back to Bethy's acting all emotional.

Which leads me to wonder who's crazier: me or my friends?

When my head's buzzing like this, the only way to calm myself is by playing music. I pick up my violin and adjust it on my shoulder and under my chin. I do love playing, but violin is not my only thing. I begin with scales, but a few repetitions is all I can muster. Boring!

I take some deep calming breaths, then begin to play a few notes of the song I composed all by my lonesome, "Edge of Yesterday." You'll probably think I'm weird, but it started out as a poem, "Ode to Leo" (da Vinci, that is):

> *There's days when I*
> *Feel like waking up*
> *Ain't gonna help me or you.*
>
> *On those days that*
> *I try to get away because*
> *Nothing gets me further from the truth.*
>
> *I travel across time and seas*
> *To rediscover you.*

You'd better be the one for me
Because I'm the one for you.

Oh Leo, Leo!

You'd better be the one for me
Because I'm the one for you.

Because that is my fantasy: to meet up in person with the great Renaissance genius! How did he get to be the "great" da Vinci, Master of All Things?

In my mind Leonardo isn't some old, dead artist; even today, people flock to see the *Mona Lisa* and study his sketches and plans for canals, gliders, and the bicycle. All of which he envisioned centuries before anyone else. I mean, who wouldn't admire a mind like that?

After pouring my heart into my playing, I feel a lot calmer. Mamma says there are studies that show now that playing music can reduce stress, and I guess I would be proof positive.

I flip through the sheet music to attempt the Mozart I know Mamma means me to practice. By now, though, my stomach is grumbling louder than the music.

Wafts of garlic, onions, and oregano assault my nose and I find myself tramping full speed down the steps.

"Me. Need. Food."

Mamma laughs.

She's standing at the stove stirring all the fresh vegetables from our garden into the simmering pot. The island is littered with tomato stems, onion peels, and the papery skins left over from the garlic cloves.

I'm salivating.

"Dinner is almost ready, *cara mia*," she says. "Please carry the salad to the table. And why don't you set it, while you're at it."

I grab an errant carrot and chomp. "But wait, Mom . . . did you hear my song?"

"I heard, *cara*." She continues stirring. Her silence tells the story.

"Geez. I thought I was playing okay. . . ."

"Um-hmm. Maybe when you're fresher . . ."

She looks at me with a critical eye, momentarily distracted from her *pomodoro* ministrations. "Charley, tell me it isn't so. Your father actually let you out of the house in those ripped jeans?"

"Ah, Mom. My friends pay a lot of money for this look," I inform her, flipping an olive into my mouth.

"Hmmph. I'll have to have a talk with that man."

"Anyway, what was wrong with my practice?"

"I don't think your heart was in the music, Carlotta! That ballad needs to meander: a slow, fluid movement."

"Slow. You mean like the way you're stirring that?" I quick dip a spoon of the luscious red sauce into my mouth.

She gives me *the look*. "More like this." She demonstrates the proper tempo with the wooden spoon. "Mellifluous."

I make a mental note to look up *mellifluous*. Good word.

"Yeah. Anyway. I'm starved. Dinner almost ready?" If I don't snack throughout the day, I get really cranky.

"Time, *cara mia*. Good things take time."

Time. Something I never have enough of. "Hey, I can stir some, if you want to do something else." I pick up an extra wooden spoon and, before she can say no, dip it back in the pot, bringing another thick red spoonful to my lips.

"Charley! Manners!" she scolds, good-humoredly swatting at me with her own dripping spoon, even as I lick mine clean. Laughing, I plunk it back into the pot.

"Hey, no double dipping!" She moves my hand, spoon attached, from the vicinity of the pot and taps me on the knuckles so they're now all gooey as the sauce begins bubbling up, boiling over onto the stove.

"Here, let me keep that from spilling over." I sneak my spoon to the pot again, but this time Mamma's quicker. She pulls the pot out of reach. "Now set the table, Carlotta!"

"Okay, okay—I surrender!" I laugh, and go to rinse said knuckles under the faucet.

Just then, Dad enters the room, completely oblivious to the whirl of activity around him. He picks up the remote, flipping channels until he lands on CNN; Mom says the initials stand for Constantly Negative News.

"Let's see if there's anything good going on in the world."

"Do we have to listen to this doom and gloom?" Mamma asks, nodding at the television.

It seems to me like the world must've been ending for at least the past 5,000 years. "Okay, here's my version of what's happening," I interject. "In today's news out of Washington, Leonardo da Vinci made his American debut at the Smithsonian!"

Dad smiles at me, a twinkle in his eye. "I am so happy we had the day together, Charley. And thanks for helping your mom. You know how stressed out she gets before a big trip!" Dad starts filling a big pitcher with water, as he turns back to the news.

Stressed out! She's not the only one.

I set out plates, forks, and glasses.

"So, Sport, what was your takeaway from the Leonardo exhibit?"

"I dunno, Dad. It was amazing. And exhausting."

"Hmm." He pauses. "Well, it's great we can take advantage of all the wonderful things our nation's capital has to offer."

"Yeah, D.C. is amazing." But I have more personal discoveries in mind.

"So *magnifico*," I begin. "Things are going well so far on my new science project. How's this sound as a thesis question: Could a Renaissance genius turn up in the twenty-first century?" It's a question I've started to ponder—because of my curiosity and passion to learn as much about, well, everything, as I can possibly soak in.

"Ahem," Dad interjects as we take our seats around the table. "Would prefer that one self-proclaimed girl 'genius' concentrate on living in her own century."

I ignore this and take my time grinding lots of pepper on top of my pasta. The spicier the better.

"You know, Leonardo da Vinci was a genius at designing technology, everything from warfare to waterways for his patrons, the de' Medicis in Florence, and Ludovico Sforza, the Duke of Milan. So, what if we could create a game to design and test out inventions in a virtual environment? Something to shorten the process."

"You still have to put in the actual work, Charley, games or no," Dad reminds me. "Discovery and invention take time."

"You know what they say, Charley," Mom interjects, "ten thousand hours . . . only ten thousand hours." This is a reference to some theory about how long it takes to really master a skill, even if you have a talent for it. "So how about you attack that Mozart before I go."

This is a problem.

"So, uh, Mom, I was thinking . . . like, I may need to take a break from music." I finally get a good angle with my fork and shove a giant twirl of pasta into my mouth. The pepper stings my eyes.

"What in the world are you talking about, Charley?"

She's caught me with a mouthful. "It's just, this project is so—me and Billy—even though Beth . . . ugh." I manage to swallow. "She's being so mean!"

The whole thing with Beth is eating away at me, even though I know I shouldn't get caught up in her drama. Still . . .

"Beth says I *do* too much."

Dad gives me a funny look. "'Knowing is not enough; we must apply. Being willing is not enough; we must do.'"

There he goes, using da Vinci's own words against me!

"Say, I have an idea, Charley!" Mom sets the salad tongs back into the wooden bowl. "There's going to be a Young People's Concert in Italy at the Pitti Palace during next year's Florentine Carnival. If you keep up with your practicing, maybe next year this time, when the symphony goes on tour . . ."

I start pacing in front of the TV. "You guys, you're not listening! Beth turned on me and was texting with this jock from school, Lex . . . but—"

"So this is about a boy?" Dad's starting to look a little embarrassed.

"Oh, you know Beth. Hormones and stuff . . ."

Dad looks suddenly desperate to change the subject. "You know what I think? I think you need a break from thinking. Maybe kick the soccer ball around with me later?"

I sit and look up briefly to see he's expecting me to say something. "Yeah. Maybe."

"Wait, do I know this Lex?" Mamma plops down in the seat across from me. She gets this mooning look on her face. "Charley, is this a boy you like? Oh, *cara* . . . !"

Dad clears his throat loudly. "Hey, I just remembered, *signora e signorina*. I have some bills waiting that are not going to pay themselves!"

Mom glowers at him as he dashes off to hide behind the big-screen TV in his office. "Now, listen, *cara mia*. There are a few things you need to know when it comes to relationships. . . ."

I feel myself blush. "Give me a break, Mother. I'm taking eighth-grade health!" I start sucking in long spaghetti strands to avoid looking her in the eye.

"Actually, I was talking about the emotional aspect: love, infatuation, highs and lows. . . ."

I really don't see what that's got to do with anything. "Anyways, Beth's become this boy-crazy . . . it's almost like she's not herself anymore."

"Oh, honey. Every girl your age experiences that magical first love. . . ."

I become more engrossed in my now-cold spaghetti.

"Please don't slurp, Carlotta. You'll make yourself sick. And you, *cara mia*—I feel like I should be here for you, honey. Your first boy—"

"What?" I suck down the last bite of spaghetti. "Mom, Lex is not my boyfriend!" I insist. Although, those deep dimples . . .

"Who then? Beth?"

"Forget it, Mom. It's no one. I'll be fine." I give her a quick kiss on the cheek and then gently rub off the red sauce my lips leave on her face. "Anyway, don't you have a plane to catch?"

She glances at the ancient grandmother clock by the door. "How did it get so late?" She frowns. "But my being away doesn't change anything. We'll FaceTime while I'm gone, *cara mia*, and if there's anything at all . . ."

I curl my lips in what I hope is a reassuring smile. "*Ciao, Mamma. Buon viaggio!* And don't worry about me and Dad while you're gone. We'll be just fine."

I kiss her on both cheeks, Italian style, and run back to my room to start my research.

IV.
TELLING TIME

I can't sleep. I get out of bed even before the alarm, since trying too hard to catch more z's only makes it worse.

Mamma normally would be making me practice violin before school. But I have too much on my mind right now to worry about it. And she's winging her way to Leonardo's city at this very moment. I figure I can take a short violin break. I feel a little weight slide off my shoulders.

The house feels off-kilter without her. Dad tries, but he doesn't really get what I'm dealing with—friends, former friends, boys who want to be, maybe, more than friends. And visions of discovery in distant times and places.

Oh, well. The good news is, while she's gone it's the only time in life I get to eat spaghetti for breakfast. Dad dishes some of Mamma's famous recipe into a cereal bowl and zaps it in the microwave.

My phone buzzes with a text from Billy: "Meet me before 1st period." (And btw, that's another thing Mamma would never allow: phones on the table at mealtime.)

I wish there had been some good way before she left to share with her my impressions from the subway ride, to see if there's any way for her to confirm what I observed from what inexplicably must be Leonardo's Florence (*Firenze*, in *Italiano*)—even if it is more than five hundred years earlier.

Too late for that now.

It's just getting light when the bus arrives at my stop at the ungodly hour of 6:30 (because they love to torture us). Beth straggles on the bus a nanosecond before the doors close and takes a seat next to me, preoccupying herself with her phone.

"What the heck happened to you the other night?" I demand.

"Never mind," she replies, snarkily. "You got me grounded, for one," she complains. "But I forgive you." She puts up her hand for a high five.

"*I* got *you* grounded? That *is* rich," I observe. Then seeing it might become World War III with us—and this is not the time or the place—I relent, slapping her hand. "And I forgive you too."

A lot of kids get on at our stop. Everyone else is scattered here and there, either half asleep or attempting to finish homework, when, seemingly out of nowhere, Beth says, "I feel sorry for the girls who are still wearing training bras at our age." Her voice is too loud. And all those kids who were just slumped over, half asleep, are suddenly staring at us.

I feel my face flushing. Beth knows I am one of those girls.

"I mean," she continues as if no one else is on the bus, "when my mom took me to Nordstrom for a fitting, the lady said, 'Oh, sweetheart! It's a wonder you can even breathe the way this old thing is pinching you!'"

Then Beth turns on this beatific smile. (Isn't that a good word, *beatific*? It means "saintly, blissful." I think of it like a Mona Lisa smile.)

She sits up straighter and practically sticks her chest under my face. One shoulder of the so-called sweater she is wearing is uncovered so you can see her bra strap, neon green and yellow evidence of her womanly sophistication. To me, it looks like someone spilled poster paints on her.

"I think Lexy will think it's sexy," she says as she sweeps her newly straightened waist-long blue-black hair back from her face like the drama queen she has become.

Sexy. I mean really.

Don't get me wrong. I worry about the same things every other normal thirteen-year-old girl does: tests, homework, boys, friends, and fitting in; why Billy was acting like a hurt puppy dog on the way home from the museum.

I also have a lot of questions without obvious answers: Who are we and why are we here? Why aren't there any modern Renaissance geniuses—like da Vinci, Michelangelo, Newton, Galileo—and who's to stop me from being the first? Asking these questions sometimes gets me in trouble.

Truth is, I *don't* fit in so well in the twenty-first century; I see myself as more of a Renaissance girl.

So Beth can boast and make goo-goo eyes all she wants; I'm making a date with Leonardo. While I listen with half an ear, my mind is running a million miles an hour. I think back to the designs I saw in Leonardo's studio. *Vitruvian Man.* The machine running on a spiral track. And the cartoons pinned to the easel with sketches. They all echoed the shape a compass traces. A shape that repeats with mathematical precision in nature—it follows the Fibonacci sequence. (This is what they call "nature's numbering system," and in Leonardo's time it was thought to represent the perfect proportions that artists and architects would strive to follow to create harmony in line and drawing.)

And one design, more than all the others, seemed to represent what I know for sure was evidence that what I experienced was *not a dream.*

I have a strong sense that what I observed demonstrates the real inner workings of Leonardo's mind: his deep curiosity not merely about what is but what *could be.* Seems to me he was doodling on dimensions that might virtually carry him into another plane of existence. Reminds me of a quote from one of Leo's notebooks: "Science is the observation of things possible, whether present or past; prescience is the knowledge of things which may come to pass."

Wouldn't it be logical that he'd try to invent something to carry him to a future time?

I wonder whether Billy and I can figure it out using current science and technology—like the whole Higgs Boson thing, finding particles that travel faster than light. Like *Star Trek* for reals: two teens on a mission *"to explore strange new worlds, to seek out new life and new civilizations, to boldly go where no man has gone before."*

The upcoming science fair's the perfect excuse.

As soon as I get to school, I run to my locker to find Billy slumped on the floor next to it.

"Hey, genius." I dial the combination on my lock. I open the locker and all my stuff tumbles out: assignment book, water bottles, just-in-case sweater, broken umbrella.

Billy scoots away to avoid falling junk and opens his mouth to say something no doubt brilliant, but before he can get out a syllable, Lex swoops in between Billy and me.

"Hey, dudes. What's up?"

I want to ask him what he thought he was doing Friday, stealing Bethy away from our sleepover. And, more importantly, why?

I admit, though, Lex is cute. And apparently a great ball player. He transferred to our school because our P.E. teacher's a baseball coach who's had players make it to the major leagues. Lex is that good, and he knows it.

But I'd never make the Lex List or even want to, so it's easy for me to be totally cool around him. "Billy and I were about to discuss the science fair, Lex. Have you chosen your project?"

Lex's face looks a little pink. "Yeah, well, y'know, it's that greenhouse marigolds-loud music thing that Bethy wants to do. I'm not all that into, y'know . . . flowers."

Bethy. So there it is. He's even stolen my nickname for her.

"Too bad we don't need any more help, *dude*," Billy bites back as he gets up off the floor.

It'd be hard to see these two guys connecting on any known frequency, but I don't get why Billy seems so bothered. A year ago I doubt he would even have noticed Lex existed. And now he almost seems bugged. Weird.

"Charley and I are all set on our—"

"Time machine," I blurt.

Billy looks surprised.

"Whoa." Lex actually sounds impressed. "That is way cooler than how noise pollution affects plants." I'm about to adjust my opinion of him one millimeter to the positive when he adds, "I could jump into the future to see what major league team's gonna draft me."

Billy saves me from saying something just as stupid—if that's even possible, which I don't think it is. "Who said anything about the future?" Billy and I always seem to be speaking the same language, even when we're not talking. "So, too bad, Lex. But—good luck with your thing."

Billy pulls me away by the elbow, and immediately a cloud of girls fills in the space around Lex. I watch the scrum briefly, and almost feel compelled to jump in. Weird.

Billy pounces on me. "What is with you, Charley?"

I assume he is wondering why I unilaterally changed our science fair topic and ignored his flying battles idea, and I can't keep my hunch from bubbling up.

"I've got a funny feeling, Billy. If we can figure out a way to prove that our way of measuring time is random, perhaps the past is going on now. Did you get that from seeing how advanced Leo's designs were for his day? What if we can find a way to send the arrows of time in reverse? I mean, it's not *impossible*, is it? Surely you can help me engineer his actual design . . . 'specially since you're the one who's into manipulating quantum particles—"

"*Virtual* quantum particles. That's a whole different thing, Charley. You can make virtual worlds behave any way you want. I know you're way into Leonardo, but this is *crazy!*"

"I am not crazy." I decide not to show Billy the gold compass. At least not yet. I can see he's not ready for the reality, even though I have seen it with my own eyes. How past meets present.

The last bells for first period have rung and we're still half a hallway away from our first-period class, Latin. I start to run and notice that Billy's still walking pretty slow.

"C'mon, Billy. Can't be late. Gotta learn the lingua franca. 'Cause I'm going back in time even if you aren't. And I need your brain to do it."

We're at the door to class, and Billy has that distracted look that means he's figuring something out.

"What's the worst that can happen? We build the thing and can't make it work. But we can still create assumptions and graph our calculations to show that, given enough scientific advancement with the Higgs Boson, neutrinos, and the positron, time travel is *possible*. I mean, after all, the Higgs shows quanta that travel faster than light, and neutrinos are particles that can move backward, right?"

I beam. "By Leonardo, I think you've got it, Billy."

"It's outrageous enough that we'll either get an A for innovation or F for overreaching."

"Time travel *is* possible. And we're the Renaissance Revival Team to prove it."

Dear Diary—The Present, for Now

Good news: Ms. Schreiber LOVED our presentation. "Great opportunity to teach us, you guys, about new theories concerning particle behavior in quantum physics. And I love the idea of pinning it to travel back to a specific time and place in Renaissance Florence." She gives us a week to come up with our plan. Seven days!

After class, Beth was acting all put out. "You think you're so ALL THAT, don't you, Charley?" Turns out Lex is doing NOTHING to help her out. Or maybe he hasn't asked her to homecoming yet. And she gave up her best friend for that?

But I can't worry about it. Billy and I got a head start on the planning at lunchtime. We're supposed to meet up at the library on Saturday to really dig in.

Today, though, I get a legit reason to skip school: Take Your Child to Work Day. Dad's gonna let me stay with him this year and not make me shuffle off with the little kids to do some stupid coloring book activity.

Instead, I'm gonna take pictures and write an article about it for the DA VINCI DAILY, our school paper.

More later, dear diary!

V.

I Spy—Dad's Work

My mom gave me the diary for my thirteenth birthday, saying I might need to share my thoughts with someone who isn't her, now that I'm a full-fledged teenager. Even though writing a blog seems much easier, I take comfort that Leonardo's handwritten notebooks have endured five hundred years. Who knows what technology will replace the laptop someday when all those millions of bits and bytes of life are lost? Hardcover journals will always be readable.

Tucking my secret-est heart's wishes under the mattress, I decide I should dress somewhat professionally since Dad and I might get called in to some sudden, top-secret meeting about national security or something. Because my father, Jerry Morton, is a scientist doing secret work for the government, but if anyone asks, he says he's a "senior managing partner." Whatever that means.

I pull on black tights and my swirly gray skirt with big patchy pockets, and a cute top. Mamma would approve. And of course, I wrestle my hair back into a ponytail. It's a conservative place, Dad's office.

"Don't you look nice, Charley!" Dad says over breakfast.

"Big day!" I'm shoveling in the eggs and washing them down with gulps of hot chocolate.

"Charley, we're not in a hurry," Dad says.

But I am. I'm already rushing off to make sure I've got everything I'll need in my backpack: calculator, tablet, energy bars, apple slices, homework—in case Dad has to work late. And I'm bringing my school's SLR camera with me to take pictures for a story about TYCTW Day in the school paper.

We arrive at his imposing marble and glass high-rise office building, and the front foyer is literally crawling with six- and seven-year-old kiddies. I march purposefully through the crowd and push the button for the elevator. I've never been allowed in his office before, but now I guess I can be trusted with top-secret info.

Dad's rambling on about how important it is for me to act mature, and how I represent him, blah, blah, blah. I snap a picture with my phone mid-rant, his face all wrinkled with worry; I can't wait to see the comments on Instagram.

" . . . and that's why this work is vital to us today, Charley," he concludes as the elevator doors open to the eighth floor, but, as we walk out, I have no idea what he's talking about. A computer printout tacked to the wall reads: "TYCTW: Welcome Children and Other Alien Life Forms."

I groan. Is that supposed to be somebody's idea of irony?

Dad marches me down the hall. Other teens float aimlessly about. I should interview a few for my article. We get to his big office with windows on two sides and a kind of little anteroom that maybe used to be for an assistant. I guess that means Dad's a grand poobah of something here.

He walks me into the small office. First thing I notice as I take off my jacket: a miniature sculpture of a weird-looking cowboy on a horse. "You can put your backpack in here, Charley," Dad says, showing me an empty desk drawer, just as his phone rings.

This closet-sized room feels even more cramped because there are mountains of papers and folders lying around. It figures that a company doing work for the government is still killing trees and leaving a paper trail. Haven't they heard about climate change?

Worse, there are no windows. I am extremely sensitive to light and noise, and there is a humming in this room that would drive me crazy in about ten nanoseconds. At first I think it's bees or flies or something, but then I realize it's the buzzing fluorescent light overhead. Geez. They haven't even switched over to LED bulbs. I'm going to have to have a talk with Dad about this.

"Sure, Bud, right there. My daughter, Charley, is here, but she can hang out for a few minutes." Dad hangs up the desk phone.

"A meeting, Dad? I'm coming with you, right?"

"Not this time, honey. It's the big boss."

We've been here five minutes, and already he's abandoning me.

"Sorry, kiddo. Now you stay out of trouble . . . and don't touch *anything*." And then he's gone.

If Dad has to abandon me, at least I can do something productive. Good thing I brought homework and my science fair research. But I can't possibly work in here.

Fortunately for me and my GPA, Dad's office is filled with light, and I can even see a park across the street through his floor-to-ceiling windows. Much better. Just have to make myself at home at his desk, which is nice and big and not at all cluttered.

I set up my tablet and spread out my library books and my notes. From deeper in my backpack, I pull out a paper bag of energy bars and a baggie of sliced apples. Fuel for the brain.

When I can't figure out how to get online using the tablet, I try Dad's computer. Except, of course, it's password protected—with many layers of encryption. I type in every password I can think of, but it's not letting me in. I'd give up on it, but Dad's showing no signs of getting back anytime soon and I've got work to do! I rummage inside the top drawer of his desk—the one with the key in the lock—and find a series of words and numbers that, as soon as I start typing them, allow me in—and lead to deeper and deeper levels of information.

(Note to Dad: It is *so* not cool to leave your passwords on a sticky note in an unlocked drawer.)

The screen flashes *Operation Firenze* three times before it goes black.

"Ha! That's funny!" I say aloud, thinking Dad posted it to remind him of Mamma's tour.

I manage to reboot and pull up a website for info on Leonardo as a scientist, even though most people today don't know this about him. It's 10:45—close to lunchtime at school—and I can feel my tummy rumbling. Dad promised we'd eat together in the cafeteria downstairs—they actually let you choose anything from dim sum to pizza, with an all-you-can-eat salad bar and frozen yogurt sundae bar.

I suck on a slice of apple to assuage my appetite and scroll through pages and pages of good information. I start typing up notes—careful to keep track of where it's coming from—when I notice something funny happening on the screen.

"Hey, wait! I'm not done yet!" It's like someone—or something—has taken over using a mad screen-share program, and things are moving around on their own: First, a Google Earth map zooms in on a close-up of Takoma Park, er, that is, home, then spins around to Florence, Italy, and closes in on old landmarks. And without my doing anything, it then maps a trajectory like the flight path of a supersonic jet—or the orbital plan for a SpaceX rocket. As I'm watching, the program transforms: It's a holographic projection of Earth, spinning wildly, but here, east to west instead of its normal rotation. A timeline at the bottom is counting back in decades, starting at this one.

It's hard to know what's happening in this augmented reality. What look to be documentary images matching history's highlights

for specific years parade above the timeline, which v
astonishing for the twentieth century, when mov
then-current events, but this timeline marches back a
decades measured out in beads on an abacus. Colors
slide in and out like a kaleidoscope.

Weird. After all, this is a normal computer. At least that's what i
thought. These images are jumping out in 3D. Does this have something to do with Dad's work?

Time and space changes continue. I have to squint my eyes to look
closer at what's happening, and I feel a hot wind, like a vortex of energy.

Somehow, the hologram starts to spin, flipping focus from the world
outside to me, and it's like the game I Spy "with my little eye" reflected
in time and out. Like, there's literally this "eye" staring out at me. Or
am I staring at it?

I hear myself shout, "My eye! *Freak me out!*"

"*Calmo!*" comes a strange voice.

mp. "Who said that?"

A black-haired god of a guy emerges from the anteroom; did I miss him sneaking in?

"Do not worry. My work here only shows the reflection of what is possible," he says, nodding at Dad's desktop. "Though your eyes are quite beautiful."

"What? That was *you* just now? Taking over the computer I mean." I'm feeling a little spooked here, and not just because the guy has one hand hidden behind his back. If I scream, will anyone hear me?

"My dad will be back any second. Who are you?"

"Kairos. Definition: just in time. And you?"

"I'm—wait. What? Definition—just in time?" It's like he knows me. "Hmm." I stare, perplexed. "Um . . . are you looking for someone?"

"*Si, signorina*: you. I am Kairos, at your *servizio*." He grins, bows, and then, with a flourish, presents me with that sculpture of the horse and rider that I saw in the outer office.

"Kairos. Looking for me." Something feels familiar about that name and the face that goes with it. I can't quite place it. "Mm-hmm. And this sculpture?"

"A maquette. From my master," Kairos says seriously, again proffering the statue. "For *madonna*."

I push it back into his hands. "*Madonna*? You've got the wrong person. I'm just your average, ordinary girl. Well, actually, somewhat above average. And you are . . . trespassing. Unless you're somebody's kid? Because you seem, well, for lack of a better word, weird."

Normally I wouldn't just insult a stranger like this, but this guy's blazing black eyes and enigmatic smile kind of freak me out. I'm clearly not thinking normally. For one thing, I'm glad I dressed for the office this morning, and I catch myself checking my hair, as if it might be doing anything other than bouncing around like it's spring-loaded.

Kairos sets the maquette on the desk and shoves his hands into his pockets, looking bashful. "You will become accustomed to me. I deal in . . . knowledge, shall we say . . . management."

Knowledge management. What a *tool*! I can hardly take him seriously.

"Well, I'm Charlotte. Charley. Are you here for Take Your Kid to Work Day?" I stick up my hand for a fist bump.

Kairos, staring, seems hesitant. "Then you are no stranger to me, and I am no goat." He doesn't fist-bump me back.

I laugh nervously. Seeing he's not fist-bumping back, I quickly shove my hand into my skirt pocket, pretending not to be embarrassed.

"Whatever."

So he doesn't belong to any of Dad's coworkers? But he does IT. And he called me *madonna*. Definitely weird.

"Oh. Well, I guess I'm 'just in time' to start my homework then." I go back to looking at the computer screen, trying to see if there's a pattern to the digital sequencing.

Kairos returns to the anteroom. Seems odd that he has no idea about TYCTW but, okay. I stop paying attention after he appears to go back to what he was doing, which apparently includes continuing to control my computer because now, in addition to the old newsreels, I see a message appear on the screen. And not random words either:

Carlotta, il tuo viaggio nel tempo è possibile.

Carlotta! Only *mia madre* calls me that. But that's insider info. So . . . who's stalking me? What if this guy is a mole? A foreign spy? As it is, he seems, well, awkward, but with smoldering good looks.

But, really, the main event is what's happening to Dad's computer. Without my doing anything, I see it zoom out from Google Earth and a holograph appears showing the blue marble that is Earth rolling on the surface of space as if it's warping the plane of space-time, apparently confirming Einstein's law of gravity. Pretty cool animation.

I point to the computer and call out, "Whoa! Did you do that?"

"It's the Qualia Rosetta," Kairos replies. "You may change the view if you prefer to see yourself." He strides back to Dad's desk and puts his hand over mine to show me how to adjust the view on the touchpad.

I feel a tiny jolt of electricity at his touch, and draw back my hand in surprise. He takes it again, and together we are guiding the Earth's movement. Zooming in on the blue ball, I land on a street view of Dad's office, showing buses and cars moving and people bustling by.

"*Ecco*," Kairos says, removing his hand. "You are doing it!"

I feel a little disappointed—after all, Google Earth has been around for years—but I click again, and it zooms back out to show the globe with a zillion stars behind it, spiraling faster and faster until it stops at the coordinates of Florence, Italy, and autozooms again to the street view. But this time the Earth view looks greener, less populous. People walking around in costume. Time moving slower.

I feel the world as I know it disappear from view. And Kairos, whoever he is, says, "You'll get used to it."

VI.
TIME IS OF THE ESSENCE

I look up at Kairos. "What am I doing? Erasing history?" As I watch, the scene morphs to show the Washington, D.C., of today. "I mean, I don't get it. We're there; we're here. . . ."

The screen shifts to night. You know those satellite photos that show clusters of cities at night taken from space? Before my eyes, I see those hot points of light dimming, shrinking, then finally winking off entirely.

"You've made some kind of virtual world?"

Kairos shrugs his shoulders. "No, a present for you: our world out of time."

A world out of time. This feels like a digital *trompe l'oeil*, or "trick of the eye." (I love this phrase, even if it's hard to pronounce. Another French art term, it describes a visual illusion used to trick the eye into perceiving a painting detail as a three-dimensional object. Note to self: Study French! Especially since France is where Leonardo lived out the end of his life.) Kairos and I are playing not only with space but also with time. A four-dimensional illusion from some fifth-dimension reality. Unreal!

"Nice game! My friend Billy would be in awe. Never seen anything like it," I say, but Kairos doesn't crack a smile.

"I'm not certain what you are saying, *signorina*. This is quite usual in my world."

"Your world. And what world is that, pray tell?"

"*Ah, signorina,* I am from *Firenze.*"

"Uh-huh. But I mean, how do you know all this? What century's information are you processing?"

"Fifteenth . . . twenty-first . . . thirty-sixth . . . what does it matter?"

Cryptic! Whoever this kid is, or wherever he's from, all's I can say is, Kairos is some kind of genius.

"*Si, signorina.* I create technology that allows one to explore yesterday. And tomorrow. Time's edges. You may want to take a look at my newest handiwork."

"Funny. Time's edges," I remark. "Everyone knows time is circular—like the seasons and days!"

"Is that so." He smiles and hands me headphones that have sensors attached like brain electrodes. They look so dorky. Even I can tell these were designed by a dude with no fashion sense.

"What're these?"

"Put them on," Kairos urges.

I'm expecting to hear music, but then something weird happens.

"*I'm expecting to hear music, but then something weird happens,*" echoes a voice—not my own—in my ear.

"Who said that?!"

"*Fa parte di me,*" comes the male voice in reply, apparently speaking Italian. But no one has said a word out loud. I look at Kairos, who's laughing and thumping his chest.

"Are you a ventriloquist?"

"It is my own invention: thought recognition," the answer comes into my mind, in English.

"You're a mind reader? You've created mind-reading software?"

"Not software, *cara* Carlotta." I swear his lips did not move. "Mindware. Something for you to keep with you through time."

Again, this thing appears to be reading my mind. "You will see it works in any language—and offers the ability to translate into your language and back."

What kind of advanced programming has Dad's office been developing? Could it be some edgy James Bond spyware thing . . . or is this part of Operation Firenze? My mind's going haywire.

Si, signorina. Operazione Firenze. A tempo . . . a suo tempo.

Whoa! Before I even have the wits to form my next telepathic question, Kairos speaks aloud again. "Well, I have completed my apprenticeship here. My master will be pleased."

"Your master? Apprentice? Is this a game?" Again, the screen seems hijacked, running 3D graphics and mathematical formulas across the face of the globe before spitting out a hologram. "How'd you do that?"

"It is nothing, Carlotta. About time, that's all. You may find it useful in balancing your passions with your studies. It is all about keeping things in proportion." Kairos then extends his hand, as if remembering his manners.

Not one to be rude, I shake, and feel a cold, hard metal object land in my hand, about the size of a key. And without even looking, I know exactly what it is.

"Oh, it's the secret handshake!" I laugh nervously.

Kairos, instead of laughing back, puts his finger to his lips and walks to Dad's window, glancing down at the street before he creeps to the doorway and pushes the door ajar to peer down the hallway.

It's then I start to worry for real about possible spies—after all, Dad does secret work—so I am careful to mask my excitement, as well as the compass, which I slip into my pocket. Pretending that I don't even notice Kairos, I nonchalantly check my email, like *nothing special has happened*, kind of like I've seen in those old *Mission: Impossible* movies.

It's then that I notice in my Google Drive a copy of the program file running on my screen and, almost simultaneously, hear a ding to my email from the address "καιρός@operationfirenze.it."

Of course, it's all Greek to me.

It's not even ten seconds before my personal super sonar detects voices coming down the hall outside Dad's office, and I hear my father telling whoever he's talking to that he's got a hot date for lunch, a.k.a. me!

I glance at the anteroom, hoping for clues here, but Kairos must've snuck out. *Oh no! If Dad sees all this . . .* I tear off the headphones, shove them in my backpack, and quickly grab a pile of newspaper to wrap up the statuette. Then I slip it, too, into my backpack and feel inside my skirt pocket to see if the metal piece—the key/compass Kairos left in my palm—is still there.

I don't have any time to inspect it now: I've got to X-out that weird computer program before Dad finds out what I've done! I barely manage to log off in time and dash back into the anteroom where he left me.

"Sport? You still here?" Dad strides into the office.

"You're back!" I force a smile.

"Yep. Earth-shattering problem solved; another crisis averted." Dad smiles and blinks (he never was able to wink). "For now, anyway. Thanks for bearing with me, Carlotta *mia*. But, today *is* Take My Daughter to Work Day. I want to hear how *your* morning went."

What can I say—that I watched, helpless, as a spy remotely seized control of his computer? That I witnessed, as if on streaming video, a hologram of Planet Earth unraveling five hundred years—or maybe raveling up fifteen thousand?

Thinking about it now, it feels like somehow the whole "infinite intelligence" program, or whatever it is, has been downloaded directly into my mind. Mindware, Kairos called it.

Again, I am at a loss for words.

Dad doesn't seem to notice. "Sorry for the interruption, but you have 100 percent of my attention for the rest of the day, Charley. What've you been up to? Did you get a chance to talk to any of the other teens? I saw that Eleanor's daughter—Marissa, I think her name is—was busy taking photos of kids visiting in other offices. Looks like they were having fun on Snaptweet, or Facechat, or one of those places you guys post stuff to these days. Maybe she'd let you use one of her pictures in your story."

He throws me my North Face jacket, still hanging over the spare chair in the anteroom, and grabs his keys. "Hungry? Let's get outta here and skip the company cafeteria for a change. What are you in the mood for—lunch is on me!"

I'm a little shaken up by all that's happened already. "Umm, Dad?" I hesitate, trying to weigh his reaction if I tell him about my close encounter with an alien culture. "I'm, uh, not too sure if I should tell you about this, but I, that is, Kai—"

"I could introduce you and Marissa, if you haven't already met her." Dad, barreling on as he does, suddenly stops and looks me in the eye. "I'm sorry. Tell me what, Charley?"

To tell or not to tell? I'm feeling a little shaky about all this. "Well, uh, you know . . . um . . . I ca-can't use photos by someone who's not from our school for my story in the school paper."

He stares at me as I stumble through this. "Um-hmm. Charley . . . is there something you're not telling me?"

Dad knows me too well. "Trouble About to Happen" is what he used to call me when I was little and getting into things I shouldn't.

Realizing trouble may actually *be* about to happen, and wishing to avert questioning, I lock eyes with him.

"Well, Dad. You of all people should understand! I mean, it's not exactly Da Vinci School news if someone from another school's taken the pictures!" With that, I slip on my jacket and pull on the straps of my backpack. The sculpture feels a little bulky, and I hope I wrapped it well enough.

Shoving my hand inside my skirt pocket, I feel the cold, hard metal and wonder again why Kairos had a golden compass just like the one I found on the subway.

It's almost like a clue in a treasure hunt. Maybe if I figure out where this piece goes, it'll unlock a new clue. I wish I had the time to play around with it, but that's gonna have to wait until I get home, alone and safe in my room.

"Is that what you're worried about?!" He slips on his own jacket. "Well, there'll still be time after lunch to snap shots of your own if you want."

"Obvi! And I've got the school's super-duper hi-tech camera with me, too," I say brightly. "Now, what about food? I'm up for spaghetti."

Dad laughs. "That's my girl! Never tired of pasta. Good thing we've got Oro Pomodoro around the corner!"

Seeing he's off the scent, I'm a little bit relieved and a lot confused. My head is spinning almost as fast as the holographic Earth tour that Kairos manufactured for me, and my stomach is talking to me. I'm hoping the upset is nothing a lunch can't cure.

"Yum, the Golden Tomato. Race ya to the elevator, Dad. I'm starving!" With that, I take off in a fast stride.

"Charley! Slow down, Sport. This is not a racetrack!" I hear his voice grow fainter, as Dad falls farther and farther behind.

Dear Diary—after TYCTW day, 2 a.m.

I tried counting sheep backward from one thousand. Put on my earbuds. Read from my history book. None of my usual tricks would keep my head from spinning. You, dear diary, are my last resort. There's something wrong about today: holographic Earth projection in a

— 50 —

clockwise rotation, years passing in reverse. The whole thing with Kairos: Where have I seen him before?

I can't quite place it . . . no, wait. The dream. Or NOT dream, as it were. That guy Leonardo was talking to . . . no wonder he looked familiar! Leo said his name, but with the accent, the roll of the R and all, I couldn't decipher it—but it was Kairos, it has to be!

Then there's the whole thing with the statue, something about the rider . . . idk. I look up MAQUETTE on Wikipedia: a French word for scale model. In Italian, MODELLO, or PLASTICO. What's with all the languages?

I know I shouldn't, but I can't resist tweeting.

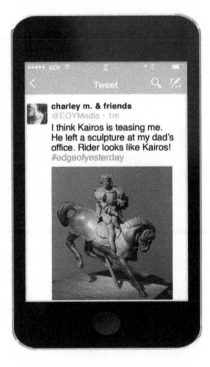

I keep thinking about where this all happened: Dad's office. Could Kairos be a spy with a cover as an IT specialist? But then, why warn me about "keeping things in proportion"? Is this compass a clue to how Kairos was able to spin the world back in time? Is he trying to help me build our machine? If so, how? And how does he even know I'm building a time machine?!

VII.
Do Your Research, People!

I've got to get Billy alone and tell him about Kairos, the game, and this mindware thing, if that's what it is. We're meeting up at the Takoma Park Public Library to launch our projects. Bethy is already there staking out a table, but there's no sign of Billy or Lex when I get there. I take a deep breath and head over to Beth's table anyway.

"Did you see Lex outside?" she asks, waving toward the door, before I've even taken off my backpack.

I still don't know what happened between them when Lex picked Beth up from my house in the middle of the other night, but I'm really not in the mood for her boy problems.

"Game day!" Lex bounds in all high energy, like a Labrador puppy. Beth rushes over to him, grabbing his hand and pulling him to our table. Billy straggles in behind, carrying an open laptop and deep in some code. That boy probably codes in his sleep.

"Billy, watch out!" I call, as he nearly clips an innocent woman waiting to check out a book.

What a motley crew we must appear: Beth preening to show off her new lacy sweater with two garish red roses crocheted suggestively, one on each side of her chest. Nothing Mamma would ever let me be caught dead in.

Not to be competitive or anything, but I decided to put on this really cute scoop-neck, swingy dress over leggings. It has pockets, of course—a must in my world, because I always seem to be accumulating bits and pieces of stuff. I found it at a neighborhood yard sale, actually. Something about the colors—black, purple, gold, and rose—that spoke to me on the hanger. Not one of those twerk-able numbers that Beth would flaunt, but it's not my usual sweats and hoodie, either. And I took my hair out of its ponytail, which I almost never do because I have so much of it, and what with all those awkward waves!

Billy's chinos are wrinkled and too short, with his thick green socks showing out from under. Lex is wearing a cool bomber jacket open over his baseball jersey (number 22, for the record).

So we're all at the same table, books and laptops open, but I'm feeling like I might jump out of my skin. I'm bursting with questions and keep biting my tongue so it doesn't spurt off what would sound to most people like I'm three-thirds of crazy. But there's only one person I can really share any of my news with: Billy. I mean, we've been in the same advanced math groups together since, like, first grade. Plus, he's the only person who consistently beats me at Words With Friends. In fact, he created a game that expands the potential to entire phrases: TeenWords. TeenWords is Billy's innovation to improve on the WWF app—with my input, of course—kind of a cross between WWF, Wheel of Fortune, and Jeopardy, with cheat codes that let you reveal entire phrases instead of single words.

He's got a strategy for winning, he says—that's a code I'm determined to hack!

Anyway, we need to talk and it can't be here, where everyone's listening. And looking. Both Beth and Lex seem to be looking at me funny today. Maybe it's the dress. I don't know whether to be flattered or annoyed. I do know that it's distracting.

"I still think I should work with you, Charley," Lex is saying, moving closer to me as Beth casts an evil eye in my direction.

"Really, Lex, Ms. Schreiber approved your project," I say, tapping my heels under the library table. "The whole sound waves–promoting–plant growth thing. Sounds awesomesauce. And you've got Bethy on your team!"

Beth chimes in. "Awesome is right. Acid rock by the Retro Pigs has gotta change the way marigolds grow in our test condition."

I bite my tongue and refrain from pointing out the well-known fact that plants respond negatively to certain sound-wave frequencies.

Beth turns puppy-dog eyes on Lex. "Right after this you and I are going shopping for our supplies, right? Seeds, trowel, watering can—"

The table jumps, and Beth edges away from it and peeks under to see my feet dancing. It takes me sheer force of will to stop them, but then somehow the pencil in my left hand starts to twirl. I quick pick it up and start to doodle on my notebook.

"Waiting, Charley . . . our research?" Billy reminds me, oblivious to the tension.

"Right," I concur. "Research."

Beth frowns like she always does when things are not going her way. I can almost read her mind, because I can tell she's trying to read mine. "Yeah. Library . . . hardware store . . . me, Lex, and the two of you. Exciting Saturday."

I feel my face starting to turn red. "Yeah, shopping—*you and Lex.*"

At this, I notice Lex drifting off toward the magazines.

"I mean," I continue despite myself, "he's so much older, and the only reason he's still in middle school is so he gets more practice time with baseball so he can play varsity next year as a freshman, and . . ."

I find my tongue running on without me, as I try to talk around why "Bethy and Lexy" feels like such a bad idea. And then the whole sleepover finale flashes before my eyes like a bad soap opera.

"What's it to you, anyway, Miss Priss?" Beth hurls back.

"What's it to me?! Do you know my mom grilled me—and I didn't know where you went. What am I supposed to tell her?"

Mrs. Meyer the librarian shoots us a look.

"There's nothing to tell," Beth hisses. "Because Lex is . . . a . . . perfect gentleman. Besides, I've got secrets on you. Like Webhead's got a crush on you!"

"Billy?"

But Billy is oblivious to this *puella-ad-puellam* wrestling match going on around him. (Translation from Latin: girl-on-girl sparring. Roughly speaking.)

"Huh? Are you guys finished arguing?" Billy says, looking up from his laptop.

That would seem to be a *no*. The uncomfortable feeling remains. I shove my hands in my pockets and there it is again, Kairos's golden compass. I squeeze it as the only solid thing in this morphing reality. Is this what Mamma was warning me about—the whole boy-girl thing?

"Yeah, well, we still have some research to do, and I'm sure what happened with Lex was perfectly innocent!"

"None of your business anyway, Charley. So get over yourself."

"Coast clear yet?" A grinning Lex comes sauntering back waving a *Sports Illustrated*. "Just checking on my fantasy team."

"Speak of the devil," I murmur.

I see the storm clouds roll off her face as Beth nods and, with great effort, pulls off a smile.

"Let's go, if we're gonna do this, Bethy—shop for dirt, or whatever." Lex plops himself in the seat behind her, peering over the computer at whatever list she's written.

Beth sticks out her tongue at me.

"That is, unless Charley and old Webhead here need some stuff, too," he continues. "Since we're off to the hardware store anyways. . . ."

Beth stamps her foot impatiently and grabs Lex's arm. "Oh, yeah, a double date. Now that would be extra fun," she says with fake enthusiasm.

"Oh, don't let me keep you two," I throw back nonchalantly. "Billy and I have stuff to work out—like the quantum physics of time travel. Right, Billy?" I turn around, expecting him to take my side.

Instead, Lex starts in. "Charley, do you really think you can pull it off? The time travel, I mean." He seems truly interested, that radiant grin dimpling his face. "I'd love to be there for the launch!"

Despite myself, I feel my face flush, and for a weird moment I'm speechless—yes, I think I will be the first human time traveler in recorded history, and yes, I think I love you, Lex. What is *wrong* with me?!

"I, er, uh . . . think you're pretty . . . I mean, you know . . . Billy, you tell," I stutter between clenched braces.

"Sure, Lex, if you think you can follow this." Billy shoots Lex a skeptical look and pulls up a website showing Leonardo's machines. "It's elementary. Leonardo da Vinci didn't know any of the elements of modern quantum physics that have turned science on its head since Newton, and then Einstein. . . ."

"Yeah, yeah," Lex cuts in. "Renaissance genius and all that. Artist and scientist and musician, I get it. But time machines! Charley, how in the universe . . . ?"

Something snaps in me as I flash back on Kairos and the computer and the metal key that is in my pocket. I've got to show Billy Kairos's key. He's the only one who can help me figure out the secrets it's meant to help us unlock.

"Yeah, well, you guys need to get to the store before dark and I just realized I need to get home. So, see ya, Beth. Bye, Lex. . . ."

"Hey, like, *ciao, bella donna*," says Lex, again bewitching me. Italian, to me, is like *la musica*.

I've got to get away from the confoundingly mesmerizing Lex. I throw Billy's windbreaker over his head and pull him by the arm toward the door of the library. "See you guys!" I yell over my shoulder and notice the others are about to disperse behind us. "C'mon, Billy."

"But Charley, we aren't done researching—"

"Shh!" I hiss.

"What's the rush? I thought we were going to spend the afternoon working here."

"Got something to show you, Billy. Top secret. You can't tell *anybody*! Promise?"

"Sure, I promise. Who'm I gonna tell anyway? Studly Lex? What a tool!"

"Aw, he's okay," I insist, although in my saner moments I agree Lex can be pretty full of himself. But when he's here and I'm here, and he's looking at me with those amber cat eyes . . . and then, there's a part of me that kinda likes that he kinda likes our project better than Beth's. Not that I'm competitive or anything.

VIII.
GIRL CAVE

C'mon, Billy. I've been working on this. Gotta show you some stuff."
We hop on our bikes and take off in the direction of the Morton family manse. The cool thing about Takoma Park in general, and my house in particular, is that it's one of those old turn-of-the-twentieth-century places with gables, turrets, front and back staircases, and hidden passages. Sears Homes, they called them, although why anyone would want to name a house after an old-school department store that started off in the catalog business is beyond me.

My room on the top floor has a random door that leads to a garret. (Picture how cool this is for a girl cave: a room on the top floor of a house, typically under a pitched roof.) It's ostensibly to store stuff; I've found old steamer trunks filled with family pictures and letters, and old legal documents. There is even a bow, I'm thinking from somebody's old violin, with all the horsehair fraying or ripped. I need to have it fixed and try it on my violin to find out if it holds any musical magic. As Mamma likes to say, the right bow can make all the difference!

I don't know if it will make my practice any easier. But if I can convince Mamma to expand on my song "Edge of Yesterday" to make it a piece for a full orchestra, it's worth it. Anyway, it's kind of a fun distraction.

As for my dad, aside from all the weird secret stuff he does, he is the coolest soccer coach a girl could ever want. So, fortunately for a modern-day Renaissance girl, what's also cool about the space under the roof is there's a long corridor where I can practice dribbling the soccer ball on days it's too wet or icy to practice outdoors, and no one's the wiser.

But mainly, whenever I need some "Charley time" to be alone or just think (which seems to be a lot lately), I have my own hideaway to escape to. Even my parents don't know how much time I've spent exploring it, and, dear reader, I'd appreciate your keeping this news only between us. As it is, they think I spend too much time by myself.

Anyway, the night after Mamma left for Italy, it helped me blow off energy because I can pace in there even when my dad is sleeping and not worry about waking him. I started gathering the stuff we might need to build the time machine model. I have gathered all my old Lego bots, some K'NEX from when I was just a kid, and of course, my trusty fortune-telling Popsicle sticks, but clearly, we're gonna have to plunder Dad's workshop in the garage for tools and more solid construction materials. After all, these will need to withstand the test of time!

But that's all beside the point: Today, Billy and I really need to focus on the logistics of building the time machine. We dump our bikes outside the garage and I tap in the code on my phone to open the door. No cars are ever allowed in Dad's personal sanctuary. In fact, I'm only allowed in on special occasions—and the science fair project counts as one.

The workbench is cluttered with the electrical and electronic innards of, respectively, an old-fashioned transistor radio, a microwave oven, and greasy bike gears and chains (did I mention that Dad fancies himself a DIY kind of guy?). He's even got my mom's old electric curlers and the flat iron she used in the fashion-backward eighties. Thank goodness those days are over!

Dad's got these two-by-fours stacked against the wall for some construction project that my mom says will never see the light of day. He's got the screws and nails in boxes on the workbench, just in case. On the floor is a two-foot-high stack of old newspapers, magazines, and recycled wrapping paper.

On metal shelves around the perimeter of the garage and overhead are racks and racks of secure computer servers and data routers flashing and beeping. I am not allowed to know what they're for or why they're here—although *Operation Firenze* does come to mind—but I do know that we had to install a whole-house backup generator just in case our power grid takes a hit, like it did for the derecho—a freak, straight-line tornado that tore through Washington a couple summers ago, toppling one ancient maple tree onto our roof and tearing out power lines for weeks.

"Whew, there're some serious electronics in here, Charley!"

"Yeah, whatever. But look at this, Billy!" I pause for dramatic effect, then plunk down my special treasure: the golden compass.

Billy picks the compass up slowly, then drops it like a hot potato—as if it might be irradiated. "Where'd you find this?"

"You wouldn't believe me if I told you," I reply.

He folds his arms across his chest and stares me down. "Try me."

"Well, I tried to tell you the other day in the hall at school. Remember?"

"You mean when you completely changed our science fair project without consulting with me? I was pretty much p.o.'d at you for that, Carlotta."

"Couldn't help it, Billy. Not after the dream I had. Well, not exactly a dream. This guy, Kairos—his name means 'just in the nick of time,' or something, in Greek."

"Wait. So you were dreaming about some Greek dude?"

I sigh. "I swear it wasn't a dream, Billy. You've gotta believe me! After your stop on the Metro, I was staring out the window and actually heard this guy talking. Only not on the Metro. There were others there, too. I mean, I could *see* where they were and it definitely was like a scene from *The Borgias*, or one of those historical TV romances. I mean, not that my mom lets me watch stuff like that. Anyway, you'll never guess who he was talking to!"

"Charley, you're making no sense. You were watching, like, a movie?"

"Well, it was like watching a movie. But it felt like I was right there listening in on their conversation. You know, almost like one of your virtual worlds."

"Yeah, well, the new game systems can seem pretty realistic. I'm working on a scenario myself that I hope to sell to—"

This is going off track. "Yeah, yeah. I know. But guess who was there, ordering Kairos around, Billy. I bet you'll never guess."

"Well, knowing you, and the fact that the Borgias were involved, I'm gonna say it was Leonardo da Vinci."

"Aw, geez, how'd you know?" I'm a little chagrined this wasn't a stumper.

"And you're sure it was Leo, 'cause . . . ?"

"It was *so obvi*, Billy. There was no mistaking that face. And the hand of an artist!"

"Uh-huh, I'm listening. So there he is, this Greek dude what's-his-name, talking to your so-called Leonardo, and you're eavesdropping on them . . . and then what?"

"Well, then we get to the Takoma Park station and we have to get off the Metro," I say.

"OMG, can't you do better'n that, Charley? I mean, really, what's the point?"

"The point is, Billy, before we got off, I heard this metal thing drop to the floor, and I'm thinking it's my lucky Susan B. Anthony dollar that's dropped out of my pocket. Even though, of course, I don't believe in luck."

"Uh-huh. Still listening. . . ."

So here I recognize the need for a dramatic climax to the story. "Ta-da!" I announce, picking up the golden compass and shoving it under his nose.

"Ew, so you picked this shiny old thing up off the floor of the Metro? Can you imagine where that's been, Charley! And you let me touch it?"

Billy's something of a germophobe, so I fish hand sanitizer out of my backpack for him. "I wiped it off, dummy."

"Yeah, well, thank goodness for that! But I don't understand what this"—and he points to the golden compass, still acting as if it might carry the Plague—"has to do with some dream about Leonardo."

"Can you let me finish? I wasn't sure myself, Billy, until—"

"Or what any of this has to do with time travel!"

"If you'd stop interrupting!" I wait to make sure he's not gonna jump in on me again. "Anyway. I wasn't sure what the compass had to do with anything until yesterday at my dad's office. There I was, minding my own business, except for stealing Dad's password so I could log on to the internet"—I realize as I'm saying this that maybe that whole password-filching business wasn't too cool, and feel a pang of guilt—"when somehow the whole thing gets a little weird. I mean, the computer, that is. It's like someone—or something—took it over completely. . . ."

"Malware? I mean, doesn't your dad do work for a spy agency, or something?"

"No comment. Anyway, there's this whole GPS thing with Google Earth, and then the Earth jerks into a counterrevolutionary spin, and there's a panorama like a newsreel through history but it went way earlier than when there were newsreels, Billy. Like centuries ago! I thought it could've been a Trojan horse at first, but then this guy around our age shows up in the adjoining office. So I ask him if he has any idea what's going on, and he starts telling me it's all cool. But he's got a bit of an accent—"

"Yeah, they're bringing in all the IT folks from somewhere else these days. You know we're always hearing about the shortage of qualified techies. Besides, I'm sure there's lots of cool stuff they work on there. So?"

"SO! Who's the guy, but Kairos!"

"Kairos? Leonardo's Kairos?"

I can't tell whether Billy believes me and is getting kind of excited, or is just totally weirded out.

"Leonardo's Kairos!" At this moment, I don't think I'll mention Kairos's mysterious reference to a provenance (a great spelling bee word that means place of origin or earliest known history) that is actually 1,500 years in the future.

"How do you know it was the same guy?" Billy asks.

"Because he gave me this!" And with that, I plunk down a separate, identical golden compass.

This time, curiosity seems to get the better of him. Billy, careful to pull his sleeves over both hands to avoid direct contact, picks up both compasses. "Whoa!" He jumps like he's been struck by lightning, and the compasses rattle to the workbench.

"What was that?"

"Heck if I know," he says, clearly rattled. "But there's something about the two together that seems to generate a jolt, like electricity."

Now it's my turn. I pick them up, but I don't feel anything except the weight of the metal. "Try it again, Billy!"

"Maybe something about my skin conducts electricity," Billy says. I think he's joking, but he's got a serious look on his face.

Tentatively, he picks the keys up again. Gingerly, he holds them this way and that, looking up to the light and then twisting to see if there's an interlocking mechanism. He finally looks straight at me, his eyes showing something that you might call reverence. "Wow."

"Wow is right."

He seems to contemplate this, trying to figure something out.

"You know what this is, don't you, Carlotta?" he says.

And there, he's got me. 'Cause frankly, in all the excitement that's happened around it, I hadn't stopped to think that it meant anything other than that there's some weird coincidence that led me to knowing we had to change the science fair project. "Um . . ."

"It's the golden ratio," Billy says, answering his own question. "From the time of Ancient Greece, artists and architects were fascinated by this number 1.61803 . . . the solution for a + b / a = a / b = Φ, the Greek letter for phi, thinking it represented the ideal of beauty, and therefore a mathematical equation for beauty. It was the perfect geometry that Leonardo worked into his paintings and sculptures. The *Vitruvian Man* with his outspread arms and legs was one of the most famous representations."

"Of course! Mathematically representing the sum of the two numbers before it in a series, like the Fibonacci."

The golden compass measures this ratio—would Leonardo have used this gauge in designing his most famous works? Perfect proportion was what the Masters of Renaissance art and architecture were in search of, after the Greeks' idea of classicism—the idea that everything in the universe held a perfect form and was designed to function . . . like clockwork.

"When I saw Leo's drawing, I suspected he would've been aiming for perfect proportion." I tap my fingers against my left temple, pondering. "But what's that got to do with time travel or weird geeks speaking Greek?"

Billy gets that look that tells me he's plunged somewhere deep inside his head, sifting through data. I swear, he's like a prototype of what humans will be like when we've evolved to a point where we're born with smart devices plugged directly inside our brain receptors. Some people might say he's on the autism spectrum, but I just think he's too darn smart for what a teen should be. Scary.

I hold the compasses up to the light myself, turning them upside down, seeing if they unlock or remain fixed in place. "I got nothin'." I tap my feet. "Well, smarty, any ideas here?"

Another long moment passes until Billy lets out a sigh. "Nope. It's like pieces to a puzzle. If what you think happened is actually true, these could be keys. . . ."

"Keys? Keys to what?"

He scrunches his forehead, as if he's struggling mentally. "I don't know. It's a great story, Charley! But you know, as a good scientist, I'm extremely skeptical of everything you've told me."

I sigh. I know it seems like I'm making things up, but I know what I know! "What if it's some kind of memory device. You know, like a thumb drive that plugs in somewhere to download data?"

"Prove it!"

"Can't yet, but we will! I just know it, Billy! I know if we keep putting our heads together . . ." I set the compasses back on the workbench as if they're rare and precious objects. Which of course they very well might be.

"We?" he asks archly. "Not me. This is too hot to handle, Charley. I mean, we're dealing with stuff that goes beyond science. Can't even pick up the compass thingies without them sparking. That's just plain dangerous, in my book."

"I'm not so sure, Billy. I mean, why you and not me?"

"Guess it's my electric personality," he says, the tremor in his voice belying his attempt at a joke.

I pick up the magnifying glass Dad keeps hanging with the tools. "To tell you the truth, I really haven't inspected them up close," I say, training the glass over one edge. I see tiny scratches in the surface. "I do see something, but it's very tiny. We might need something like an electron microscope to look closer. But writing or pictures aren't the only way information can be transferred, especially if we're talking about magnetic fields or digitized information."

"Gimme that magnifying glass. Bet I could figure out the code with the right magnification. . . ." The mention of an electron microscope seems to have gotten his attention. "And gold does conduct electricity. But if this, y'know, Kairos guy comes from Leonardo's time, they certainly wouldn't have had electrici—" Billy's brain is terminally in detailed logical processing mode, whereas I believe in taking giant leaps.

"Lightning strikes. Auroras. Solar flares. Compasses—electromagnetism. I dunno, but it's worth thinking about, Billy."

"Okay, I gotcha." He pulls his jacket off and drops it in the corner. "But would that be enough to power a time machine, back in the day?"

"Of course, Leo wrote backward and in Latin, so deciphering his notes is like a puzzle. But it's obvious to anyone with eyes that the wigwam-with-tracks thingy showed his conception of a time machine! Only Leo wouldn't have had the science or technology to make it work."

Billy looks puzzled, but to me, this is a no-brainer. "Solar power. You know, if we rigged up a sort of solar battery, we could test it out!"

"Yep, that Italian sun could get pretty strong, I'm thinking," Billy says, scratching his chin. "Then, that sketch of Leonardo's in the Codex—we're gonna have to build something portable for the science fair or no one would believe it. And even then there's really no way in Dante's *Inferno* that the thing could actually work."

"Yeah, well, I figure we've got all this lumber here, and servomotors from Lego bots, and Dad'll hardly notice if a couple pieces are missing."

"Hmm. I guess we start first with a scale model, okay? See if it's even feasible?" Immediately, I see Billy's eyes scanning the room for parts.

"OMG, that's it! A scale model! Wait here a minute, Billy."

I run up to my room. I fling open the door to the eaves and duck under the low doorway. It's dark in there, so I shine the flashlight from my phone and there it is—the maquette, still wrapped up in the newspaper I shoved it in at Dad's office. I gather it up in my arms and walk down the stairs, slowly this time, to show Billy.

When I get back, I can see Billy's found my piles of fortune-telling Popsicle sticks, and he's glue-gunning them into a model that is starting to take the shape of a scale replica of a tepee, about three feet tall.

"Ta-da!" he grins, waving the glue gun a little too close to my hair. I duck just in time, but Billy hardly notices in his jubilation. "Just like old Leo's model."

As you might imagine, I am not all that pleased to see my fortune piling up on the garage floor. On the other hand, what better way to see what's in my future than a trip to the past?

Then Billy waves my mom's old flat iron in the air—he's created a solar panel, with colored glass panels wrapped in foil.

"Wow! How'd you do that?" I ask, examining his handiwork. It's a crude model. "Will it do the trick, do you think?"

"Been working on this invention for a while, Charley," he explains. "I want to make my virtual worlds operable under remote conditions, you see, so you need to use whatever power source is at hand. Gotta think ahead, here, right?"

"Oh, that. Minor problem," I reply. And then I step back to get a better view of what we've got.

Here, then, by the grace of who knows what, stands the frame of a mini-machine that follows Leonardo's design, only out of Lego bots and K'NEX and powered by the tablet and a makeshift solar battery.

Forgetting in my excitement why showing Billy the maquette was so urgent, I squirrel away my newspaper-wrapped treasure. We work together on assembling a couple of panels.

"This is a lot harder than it looks," I observe, trying to keep the foil from crinkling. "Maybe we should've kept the flat iron free to smooth it out," I say, ripping the foil for the third time in an attempt to flatten it. I'm not the master of model-building—did I mention my two left thumbs?

"Um-hmm," Billy says distractedly. But I see he's mastered the whole foil-wrap process.

"This is gonna be so cool, Billy! I mean, what if this 'machine' actually could work—activating time travel, that is?"

"Work? Who said anything about the thing working, Charley? Because I'm not so sure, with what we have here, we could actually operate your so-called time machine," Billy says, rubbing his chin. "After all, it is material, and no one's exactly shown that matter can dematerialize here and rematerialize somewhere else. Much less in another time zone!"

"Wait, who was it who told me time travel is *not impossible.* So if it might be, and Leonardo actually planned his design for this purpose, and we could . . ." I'm practically jumping up and down with excitement.

Billy frowns and rubs his hand like the earlier electrical shock still stings. "Sorry to burst your bubble, Charley. It's really pretty unlikely the thing will work. I mean, why should we assume this compass is for anything but drawing and measuring geometric shapes?"

"But we've got the computer program on my tablet—you know, Kairos's demo. That, to me, is evidence of the possibility that we really have something here. And if . . ."

"Um, I don't think it's . . . er, what I know is . . . actually . . . I'm sorry, Charley. Fun figuring out what Leonardo might have been thinking about and all, but when it comes to actually working . . . ? Did it ever occur to you that you might just have dreamed all this?"

My heart sinks. His words echo what everyone's always telling me—"What a wild imagination you have, Charley!"—just before they pat me on the head and tell me to "come down to earth." Now, even my science fair partner thinks I'm just making this up!

"You're abandoning me?!" I duck my head so he can't see my fallen face.

My disappointment must be showing, 'cause Billy puts an awkward hand around my shoulder.

"Yeah, um, well. Nothing personal, Charley. But, you know, I've got the future of flight on my mind. If not the history of winged warfare, maybe I can do something with solar-powered flight."

He holds up the makeshift solar battery and attempts a grin. "And I'm betting that requires less magic than your so-called time machine."

"Whoa. Way to be a doubter, Billy."

"Plan B. That's all." With that, Billy pulls on his jacket.

"Hmmph. Suit yourself, traitor. And don't let the door hit you on the way out!" I have a hand on the garage door opener, hearing the motor grind as the door raises. I turn my back so he can't see me as I dissolve into tears.

IX.
ON MY OWN AGAIN

Billy's defection hurts. It's lonely trying to do this on my own. But there's no question about my continuing this project, Billy or no. I pull a clean-ish tissue out of my pocket to wipe my eyes and try to pull myself together. "It's gonna work, Webhead. You'll see. And then you'll be sorry you abandoned me for your stupid model plane!"

Nothing to do but get to work, I think with a sigh. Hoping some flash of inspiration will come in Billy's absence, I walk around the little model, cradling my left elbow with my right hand and drawing my left fist to my chin. It's a pose I've seen Dad take countless times when he's puzzling through how to fix something.

Could I mock up an electric mechanism using Lego bot servo-motors—the motors that turn Legos into functioning robots ('cause you never know when an extra robotic arm might come in handy)? But the thing's going to need an extreme power source if there's any hope of it ever working.

Getting this thing to operate will require some involved mathematical calculations—something Billy's brilliant at. I sigh, wondering if I would be selling myself short if I tried to win him back on the case. In the meantime, Kairos's demonstration aside, I vow to read up on how scientists proved that the Higgs Boson can travel faster than light—undoubtedly a key to unraveling this mystery. That, or quantum entanglement—the idea that two particles separated by miles still influence one another. Which leads me to think of the "many worlds" idea—that we live in a multiverse—and then, what the heck could my traveling back in time do to the world as I know it?

All of which requires a more advanced class than middle school physical sciences.

I lose track of time, so engrossed in studying the various theories post-Einstein: Wormholes. String theory. Entanglement. All theoretical.

Suddenly, I hear a racket that seems to be coming from the driveway and worry that Dad might be home early from work.

"Dad? Is that you?" No answer. I must be hearing things. I think the craziness is finally getting to me. *Just keep focusing, Charley*, I tell myself.

More clatter—this time, right outside the garage door. I jump. "Who's there?!" I take newspapers from the recycling pile and throw them over the time machine. Just in case.

"Charley? Hey, open up. It's me, Billy!"

"Billy?" I punch in the house code, then watch as the door slowly creaks open. I'm surprised to see it's almost dark out now.

"Well, look who's here. Giving up on your solar-powered plane so soon?"

"Thought you might be missing me," he says with a smile as wide as the Cheshire Cat. "Or this." Germs apparently forgotten, he holds up the second golden compass.

"Yikes!" I grab it away.

"I must've accidentally thrown it in my backpack when I left."

"Oh. Well. Thanks. I guess." I look at him, waiting.

"So, like, I've been thinking more about our project, and, um . . ."

"*Our* project?"

"Um, yeah. I'm sorry, Charley. It's just that, well, you know. It's so *bold* and everything. I got scared."

I'm grateful that he would admit this. It's kind of what I've been feeling too, ever since Kairos . . . and the reverse rotation of the planet. . . .

"I know . . . ," I say hesitantly, "'cause, um . . . I'm scared, too. Thanks, Billy. For believing in me, I mean."

He flashes a smile of relief and sticks out his hand. "Friends again?"

I nod, taking his hand in both of mine. There is a depth of genuine feeling in his eyes I never noticed before. It feels good. "Friends."

"So before I left, you mentioned something from Kairos on your tablet. Some kind of demo?"

"Oh, right! Kairos's mathematical formula—the Qualia Rosetta, he called it. He sent it to my Google Drive." I begin pacing as I try to explain. "Remember how we hypothesized there being a Rosetta stone to decipher energy frequencies with parallel or past worlds? What we were missing was the key in mathematical terms, but this—this could unlock a whole deeper dimension—a holographic universe."

I pull the program up on my tablet, and miraculously, the Earth begins its counterrevolutionary spin. Billy stares, transfixed for a moment.

"Yes, as quantum mechanics suggests . . . but I fail to see . . . ," Billy stutters, cogs turning in his brain. He can be maddeningly slow to see what's right in front of him.

"Like you said—the formula, the golden compasses—it's all got to come together to solve the puzzle!" I point to my head. "No need to say it, Billy: 'Charley—you're a genius!'"

"Charley, focus! It's gonna take more than genius here. First of all, the curve of space-time, if the thing actually works—which I strongly doubt because of the qualia of that Rosetta—means displacing the actual past with a hypothetical alternative. Which is highly unlikely. Second, translating the sensation from the subjective experience to objective reality based on the permeability of time is only known to be applicable at the sub-molecular level. Since you are definitely a material girl, not some quantum particle, you'd have to turn into energy before that could happen.

"And third, you couldn't go back to a past that's actually happened without changing today. You know, the paradox problem."

"C'mon, Billy. They do it all the time on *Star Trek*. Besides, I promise to just go look—not interfere with history."

"Hmm. Right. Even assuming all those things fall into place, we'd still need an accurate calculation for place. So what would the coordinates be for Florence, Italy, adjusted to the more than five-hundred-year time difference? Would something like continental drift or seismic activity have to be factored in?" Billy looks at the compasses again, like the answer must be with them. "Of course, even if your formula solves for all that, Charley—and that's a pretty big *if*—you'd still have a language problem—like, Latin? Italian? As in, how would you and Leonardo even be speaking the same language?"

Whoa . . . that reminds me! "Wait here a minute, Billy." I dash back up to my room. "The translating headphones—they're here somewhere!" I mumble. I rummage on my desk and under the covers of my bed, then under the mattress where I am reassured to feel my journal, still hidden where I left it.

"Charley?" I hear Billy call after me.

I was able to examine the headphones a little closer in my room after TYCTW Day. Not to let the sixties' retro look fool you, friends; this high-tech piece of technology has an extremely sensitive microprocessor built in. The wireless headband part has electrodes that can attach to pick up on mental brain waves—EEG recorders to measure voltaic fluctuation from the ionic current flow that can be decoded through the microchip. This, I deduced, allowed Kairos to pick up on my thoughts.

To give these ugly things a slightly more fashion-forward look, I crafted little ear covers out of fake fur from an old pair of earmuffs I found in the attic. Purple fuzz. I hid the purple headphones in case they should weirdly start giving instructions, like Kairos did to me at Dad's office. If someone else were to hear them . . . !

I shine the flashlight from my cell phone into the dark under my bed, around the dust bunnies. I punch away my old soccer ball, disregard my old soccer cleats that are too small (but that I can't bear to throw away yet because I wore them when we won the county soccer tournament for our age group), and finally get to three plastic storage containers with my winter clothes and sweaters.

I tucked the trippy translation device in one of the boxes for safekeeping. I inch my way under the bed and flip open the first box. "Achoo!" I sneeze, the dust tickling my nose. I feel the fuzzy covering of the headphones, grab them without looking, and run back downstairs to find Billy hard at work.

I can see he's set up his laptop and is trying to figure out how to input GPS coordinates using Kairos's formula, now downloaded from my Google Drive.

He grins. "Ta-da—exhibit A here. So, who's the genius now?" On his computer screen is a map of the globe with a spinning cyclone moving across its face.

"Wow, nice weather map," I observe. "Looks like we may be in for a bit of turbulence soon."

"I'll say!" Billy says, a doubtful look on his face. "We may be unleashing storm technology, for all we know, Charley. Are you sure you really want to do this?"

"Sure," I say, handing him the translator. "So, then, how do you like this for exhibit B? Put 'em on!"

Billy makes a face as his fingers brush the fuzz. "Ugh!" He's kind of tactually challenged. That means he's extremely sensitive to texture against his skin. "What are these?" He tentatively puts the muffs over his ears.

So who's the genius now? I repeat silently. I purse my lips.

"Whoa, what's that? 'So who's the genius now?' How did you, like, say—wait. Are you mocking me?"

I laugh. "Nope. This is advanced technology—mindware! Translator, I call it." I see Billy's about to geek out over this techno marvel, and realize it could take us down a whole new path.

"Yeah, pretty cool, no?" I need to get him back to matters at hand. "But wait'll you see this, Billy—exhibit C." I pull out the package I'd brought down earlier, before Billy's defection. "I think it's another clue."

He looks down and up, up and down. Then he eyes me. "Yeah. Well. Big clue! Don't tell me the newspaper was printed in 1492, Charley. I mean, Florence back in the day was not actually getting *The Washington Post!*"

"Huh? Oh, no." I quickly rip away the newspaper to unveil Horse and Rider. "Whaddaya think?"

Billy raises one eyebrow. Then he bends in to get a closer look. "Where'd you get this, Charley? And don't tell me Kairos."

"Then what am I supposed to say?"

"CHARLEY! HEY, YOU GUYS. I SEE YOUR BIKES, SO I KNOW YOU MUST BE HERE. CAN YOU LET ME IN?"

"Lex!" I feel butterflies all over. Lex is the last person I need to see right now.

"Don't answer, Charley," Billy whispers. "Pretend we're not here. We can't let *anyone* see this."

But Lex is pounding on the garage door now. "Charley? I need to see you."

How can I ignore him? "Let me go out and talk to him a minute, Billy." I sweep the flotsam and jetsam on the worktable into my backpack, including Billy's solar battery, while Billy begins a ploddingly slow job of wrapping up the tepee frame. I quick rewrap the maquette and stick that in my backpack, too.

"Hurry, Billy!" I hiss through my teeth, as I do a quick final check of the things that might give us away. I'm about to breathe a sigh of relief when I spy the laptop, which Billy had left propped up against the tool-box, still playing Kairos's holographic program. I snap it shut and shove it behind the toolbox. Even for a guy as clueless as Lex, that stunning, backward-spinning Earth would've been a dead giveaway.

X.

Quo Vadis

I allow the garage door to open up enough to slip under. There stands Lex, bomber jacket and all. "Hey, Lex. Thought you and Beth were doing your supply shopping."

He's trying to look cool while sneaking a peek under the garage door. I dance around enough to block his view. "Oh, yeah. We finished." He tries to walk past me.

I stand firm and cross my arms. Need to stall so Billy can get things under wraps, so to speak. But it's beginning to drizzle.

"Um, Charley? Could we go inside maybe? I don't mind getting wet, but Coach, well, he might get upset if I catch a cold or something."

"GO INSIDE, YOU SAY?" I repeat, using my outdoors voice so hopefully Billy can hear. "Why, sure, Lex! Uh, yeah. Well, this door's a little tricky. Let me duck under so I can get the door to open all the way."

Even as I'm congratulating myself for fast thinking, Lex reaches over and grabs my phone. "Is this how you open the garage door, Charley? 'Cause we've got the same security system."

He randomly starts pushing digits. "Hey, Lex, cut it out! You're gonna mess everything up. It's extremely sensitive!" I have to pull the phone away and enter in the proper code pronto before he sets off the alarm. I know Billy's gonna be scrambling to clear away the evidence. Finally, the motor kicks in to raise the door, and there's Billy, sitting in one of the camp chairs my parents bring to my soccer games.

"Why, here's ol' four-eyes! Whaddaya know!"

"Hey, Lex," Billy says, cool as a cucumber. "Didn't expect to see you here. Wassup?"

But Lex is too busy looking around. "Wow, cool stuff here!"

"Yeah," I say. "My dad's workshop. And we were, uh, working on . . ." I scan the garage but don't see any evidence of our recent engineering marvels.

"Background research," Billy interjects. "Y'know, there's a lot of planning that goes into this project, Lex." Billy stands and hits Lex playfully on the arm, then, obviously hurt from Lex's dieselness, rubs his right fist in his left palm.

"So you said back at the library, Webhead," Lex says, unfazed. "So?"

"For example," Billy continues, "did you know that speculation about the faster-than-light properties of the heretofore merely theoretical Higgs Boson has recently been confirmed by secondary experiments in the CERN labs in Switzerland?" Billy gesticulates wildly, pointing and punching the air all around as he orates. Following his hands, I see a large "package" wrapped in balled-up newspaper, perched on the edge of the workbench. Far from being hidden, it takes up a good portion of what empty space there is on that table. In fact, it looks like it might topple over at any moment. Not too subtle!

Billy meets my eyes and nods, never pausing as he goes into particle physics' confirmation of the positron—a charged particle that moves seemingly backward in time.

I can see Lex's eyes glazing over. "TMI," he mutters.

"Too much information is exactly the point, Lex!" I confirm, defending my friend. "So that's where we are on all this!"

Lex seems to rouse himself. "Yeah, but the time machine—where is it?"

"Oh, that!" I say. "Not quite at that stage yet. Why do you ask?"

"I can't talk about it in front of old Webhead here, Charley."

Billy does a big fake stretch. "Oh, will you look at the time! My mom expected me home a half-hour ago."

"But, Billy—"

"It's okay, Charley. We'll work on this again tomorrow." Billy hurries out to his bike, me following.

"Billy," I whisper, "the, um, y'know . . . the key. Where . . . ?"

"Oh, got that puzzle all wrapped up, Charley," Billy says. "No worries! What you need to worry about is that guy in there." He pauses for effect. "I don't trust him. Not one minute."

"Wait. Are you suggesting that Lex would try to steal our science fair project? Because I am pretty sure Lex is not smart enough to even think of that, much less know what to do with it. Unless Beth put him up to spying on us."

"You know, dumb jocks are always looking for the easy A," Billy adds, with a note in his voice that, if I didn't know him better, I might think sounded a bit like jealousy.

Once again I've let my science brain get scrambled by emotion and hormones. It's like I hear Beth's voice in my ear saying mean things about me. In self-defense, I throw my one-two punch in Billy's direction.

"Ha, ha, don't make me laugh."

"Do you want me to stick around? You know, for protection?"

"Billy! Nothing's gonna happen. Besides"—I wave my phone under his nose—"I've got you on speed dial."

Billy looks toward Lex, who's swinging my old mini lacrosse stick like it's a bat. "If you're sure—"

I nod, showing a level of control I am not feeling. "I got this, pal." I shove the phone into my pocket.

With that, he jumps on his bike and pedals off into the rain, leaving me to return to a suddenly earnest-looking Lex. He's set down my little lax stick, and now that it's the two of us, I can again feel my heart beating out of my chest.

Lex hits the garage door button, and as the door lowers he puts his arm around my shoulder, pulling me closer. "You look really nice today, Charley."

I pull away, but I'm sure I'm blushing. "Whew, a little claustrophobic in here, don'tcha think?" I ask rhetorically. I look around anxiously, wishing there were at least a window to crack open. All I see are the metal shelves filled with whirring and buzzing machines.

"I can see why you'd wanna make the time machine here, Charley," Lex muses. "All this electronics gear. Bet you could power up some amazing invention using all the juice in here!"

His eyes dart around the garage until they land on the awkward mountain of newspaper on the edge of the workbench that's hiding our contraption. "Hey, wonder if your dad ever needs somebody to help organize his tools and stuff for a couple extra bucks? I do that at home for my dad—when he's around and sober, that is, which isn't all that often. I'm trying to earn extra cash to go to baseball camp this summer. All the college scouts visit to scope out talent. It could mean a scholarship for me someday."

"Is that the big secret, Lex, that you couldn't talk about in front of Billy?"

"Well, that, and . . ."

"And?"

"Well, you know I kind of like you, Charley. Right?"

I blush. No. I didn't. Not really. "Um, well, thanks, Lex. I guess."

"And that I think you're *way* cooler than Beth. I mean, she's trying *way too hard*. We stopped for pizza after, and she picked this booth, then patted next to her for me to sit down, like this . . ." He pulls my hand toward the workbench as if to sit me down beside him.

"You do know that Beth's been my best friend since second grade, right, Lex? And that I really would never—"

But he keeps on talking.

"Not you, though." Lex leans in again, drawing a hand down my shoulder. "I like a girl who plays it cool." I find myself shivering despite the heat radiating from Lex's body.

I flinch, almost tripping over the workbench, but somehow he's holding my hand, somehow he's keeping me from falling. Somehow, I'm not quite in my right mind. "Well, that's Beth!" I say brightly, slinging my backpack over my shoulders as a signal that I've got someplace else to be.

Lex doesn't seem to notice. "But what I like about you, Charley, is you'd never confuse a guy."

"Umm, thanks?" I say, not knowing whether to take that as a compliment. I pull my hand away, hard this time.

At this, Lex seems momentarily confused. "I don't know what to make of it. Me, thinking a brainiac could be . . . I dunno. But maybe we could, y'know, hang out together sometime?" Lex continues, staring at the floor but kinda pacing off across the garage, as if measuring his steps helps him think.

"Yeah, well, maybe . . ." I'm skating on thin ice as the conversation takes yet another unexpected turn. I've really never been asked out on a date with a guy by myself before. "Y'know, we could all go out for pizza or something. All of us. I mean you, me—the whole posse."

Lex is wearing a dangerous look, eyes glinting yellow under the garage lights. Think fast, Charley!

"Say, Lex, would you, um, like to see how we're building a model of the time machine?" When I see him looking even hungrier at this suggestion, I immediately regret it. I could kick myself as I realize, where Lex is concerned, my mind and my mouth have gotten totally disconnected. It's that whole overly emotional adolescent brain thing. So annoying!

But I need to get back on terra firma, so I start explaining how time is relative, and what Einstein's predictions about light speed really mean, and I see Lex isn't paying any attention. So I'm pulling more miscellaneous stuff out from the mess I shoved into my backpack to show Lex things I'd need if I were really going to time travel: Billy's solar battery rigged up with those glass panels; an old rabbit-ear antenna, a relic from the ancient television Dad—Mr. Fixit—somehow keeps in working order out here; plus juice boxes, gummy worms, and energy bars to sustain the worldly time traveler until she gets to her destination . . . usual stuff. I do an instinctual pat of my pocket to check for my cell phone and toss that in, too—to communicate with the modern world via Facebook and Snapchat.

"So, you see, Lex, it would be important to ensure some way to stay in touch with the present. Er, I mean, the future. From the past, which is, or was . . . er, um, you know. Now."

He seems to regain his footing, too. "Sure. Cool, Charley. But could a guy really see into the future? Or, like, actually *go* there?"

"Whaddaya mean, Lex?" I know well enough what he means, but if I've learned one thing from Beth about guys, it's that they love to show off. So I'm down with playing the game. For now, anyway. I shove most of the stuff back into my backpack and sling it over one shoulder.

"So, say a guy—you know, a really talented young baseball star— wants to travel into the future to see what MLB team is going to draft him because, you know, my pitching strategy might depend on the team lineup . . ."

"Well, sure. You'd want to know the whole future lineup," I say. "'Cause you know Bryce Harper's gonna be really *old* by the time you come up to the majors!"

Lex laughs. "Bryce Harper—old! Oh, Charley-o! You are *soo* cute when you talk baseball."

I don't say it, but I'm thinking he's pretty cute when he tries to be serious. He's showing that dimpled grin that has all the eighth-grade girls in a swoon. And I am not immune to a beautiful face.

"So, like, Charley, what if your so-called time machine could send me to my own future? Would you do that for Lexy?"

I feel myself about to crack up, and it's a big effort to arrange my mouth into anything other than a huge grin. I'm not sure how it looks from the outside, but I must look mad, or something, because Lex says, "C'mon, Charley. I know you could do this if you wanted to! Pretty please, with sugar on top?" He walks me into a big bear hug, backpack and all. It's pretty awkward. This is the first time a boy has ever done that to me, and it's all I can do not to melt. I know all about how physical contact releases the hormone oxytocin, which is all about bonding, like for moms and their babies.

But then the adrenaline kicks in. What if Dad comes home and finds me in here alone with a boy he's never met? The parentals are already suspicious of Lex, ever since I told them how Bethy has been practically throwing herself at him in the lunchroom at school. Like I care.

Except now, Lex is suddenly hanging all over *me*! But being a girl of science, as I like to think I am, I ponder Lex's question seriously and duck out of the hug.

"Well, it's purely speculative at this point, of course, Lex, but I suppose theoretically . . ." Immediately, I could kick myself: Charley, again with the science-speak! And obviously, that would make Lex bored. Restless, he starts to look around, poking into Top Secret spaces.

"Hey!" I yell. "That's my dad's—" He's touching all the blinking servers, until he gets distracted again by something flashing and spinning. Only then do I notice my tablet is still out. I gasp as I realize I've left the mysterious program running from when I was showing Billy. Kairos's globe is making its counterrevolutionary spin in holographic space.

I run and grab it but not fast enough to hide the screen.

"Whoa, awesome animation! Is this the thing, by any chance?" Lex asks.

"Oh, that's nothing. One of Billy's virtual worlds. It's in beta; you can't really go there yet. You know Billy. . . ."

I shove the tablet in my backpack and my hands close in on something fuzzy. Kairos's translator! Lex looks impressed when I model them.

"Pretty cool invention for a geek," he says. "What's this, then?" His pacing has taken him to the workbench, and I realize what a slapdash job Billy did on the wrapping.

Before I can stop him, Lex is tearing away the paper.

I throw myself toward the workbench, trying to protect the half-exposed model, and I brush against Lex. The golden keys clatter, metal against metal. I feel a shock—like a lightning bolt running through my body—then Lex's lips on mine.

And the world is spinning, spinning, spinning. . . .

XI.
OUT OF TIME

"Lex? Where'd you go?" My scream comes out so soft that even I can't hear it. No one answers. In fact, the silence is so deafening, I get the feeling I'm spinning in a vacuum. I struggle to hold my head together because whatever is happening is moving so fast, I can't process it. I feel like Alice falling down the rabbit hole.

Fleeting images swirl in and out of my brain as fast as synapses firing. Lex. Popsicle sticks. Rabbit ears. I manage to grab my head and realize I still have the headphones on. My mind registers that "Charley has left the building," but how—or why—is beyond me at the moment.

A cosmic Kiss and Ride. All I can say, from whatever reality I am currently living in, is . . . WOW. I'm inside something like a hologram on an IMAX screen with surround sound and maybe even odor, 'cause something like smoke is wafting through my nostrils. Colors slide in and out like a kaleidoscope behind my closed eyelids.

Then the sound starts, a kind of song—a *whooshing*, like whale music, only the pitch goes high-low, high-low, like drilling from the heights to the depths, or listening to Rachmaninoff. I remember my music teacher said that everything in the universe has a song. I didn't believe him.

I feel like I'm on a roller coaster entering a downhill spiral at g-force. Well, I do love a good roller coaster. What is it that Mom and Dad always say? If I stop trying so hard to figure things out, I might even enjoy myself. Although I'll bet they never had anything like this crazy ride in mind.

But before I can give peace a chance, I actually feel the gravitational pressure change, as a desert-hot wind fires up around me. Immediately I feel like I'm in a popcorn bag being nuked, wrapped in a warm vortex of energy—sort of like a clothes dryer on a super-powered spin cycle. My science brain thinks this must be what it's like to ride in a cyclotron with the power to separate out all the particles in my body and reassemble them. And I'm floating above and watching myself from below all at the same time. Meanwhile, there's this familiar music echoing around me: my own composition, "Edge of Yesterday."

Maybe I'm going crazy! It's like I'm suspended in time while the hologram around me spirals. Or I do. It's hard to be certain which, and it's even harder to get my bearings. Which way is the floor? Where is the roof of the garage? Am I falling or floating or spinning in place? If only I could reach out and grab something familiar. Or grab anything at all. It'd be so sick, if it wasn't so scary.

I try to open my eyes, but some force—and it definitely feels gravitational—is too intense.

Sooo dizzy. Pretty sure I'm going to hurl.

Then—SONIC BOOM! And—

Dead. Silence.

And stillness. I seem to have stopped moving.

I open one eye—feels like the other may be swollen shut—and see nothing. It's as dark as midnight, and as I try to look around, my next sensation is that of pain. A lot of pains. All over. Maybe I passed out and hit the ground? Hit it hard, apparently. My tush is wedged against something rigid and lumpy, my ankle is throbbing, and both of my arms hurt so much that for the moment I decide not to try to move the right one to feel if the left one might be broken.

My nose, however, is operating fine. Smoke fills the air with the scent of rotten eggs, reminding me of my chem lab experiment last year that backfired, filling the room with acrid sulfur. I hope I haven't set my dad's man cave on fire, but as I try to clear the fog that clouds my brain, I look around and see not flames, but mud, gravel, and weeds. I'm wedged up against a crop of rocks. I'm outside? Why doesn't that compute? Did I somehow create a blast that blew me through the garage door? Or is my memory short-circuiting?

What. A. Trip. I feel like when I got off the Tilt-a-Whirl at Six Flags: dizzy and a little sick to my stomach. But already the strange landscape is distracting me from throwing up. What I see looks like no theme park.

But where am I? The last thing I can recall, we were with the gang at the library, then I was riding my bike home so I could show Billy . . .

but then there was Lex and a lip-lock before the world started spinning and everything went black. Could an unforeseen kiss and an awkward misstep be the unpredictable trigger that unraveled everything?

OMG, this is the model! That a quantum reaction might be set off by surprising and uncontrolled emotion and could lead to . . . yes! Personal physical displacement in space and time, a.k.a. TIME TRAVEL!

My brain's in overdrive—a zillion thoughts. The science fair project. Time shift. Something feels familiar, and my mind flashes back to the dream on the Metro.

The air cracks with an explosion that sounds like another sonic boom; a sudden flash of light and a black metal ball punctures the air, hurtling in my direction, then another, and another.

Whoa, here's another experience entirely! Was that cannon fire?

No time to think: I quick squeeze up against the rocks. Another bounce close at hand: I hold my breath, shut my one good eye, tear my headphones off, stick fingers in my ears, and count down: five-four-three-two . . .

Nothing.

Bounce, and . . . nothing.

Duds. I quick exhale, praying I won't start coughing. Then I peek around to try to identify the attacker, but it's too dark. Things seem to have quieted down; perhaps whoever it is has stopped to reload. I wait.

When I hear nothing more, I cautiously remove my fingers from my ears and check around me to make sure there are no more incoming missiles. Luckily, it seems whatever battleground I've descended on is quiet for now.

But then I hear the sound of boots running and a deep, muffled voice that rises to a shout as it comes closer and closer.

"Mwah-mwah, mwah-mwah, mwah!" comes a deep, melodious voice. I couldn't tell you what he's saying to save my life. Where in the world am I? What's going on?

Fear grips me. I spin, trying to control my trembling as I feel him closing in on my poor excuse for a hiding place. And then he is standing in front of me.

XII.
Was That a Selfie?

I know that face.

It's him.

Leonardo.

"Hey, ragazzo! Ti sei fatto male?"

"Huh?"

Leonardo da Vinci holds a torch that illuminates a small circle around us. He offers a hand to help me up. He seems to be frustrated, but it's all mumbo-jumbo. *"Palle da cannone colpire il bersaglio, e . . . lui."*

"Uh, me?" I point at myself. If I'm really where I think I am, first-year Italian is not quite going to cut it. I blink hard, willing my brain to translate.

Leonardo is pointing at me, too. *"Si. Chi c'è qui?"* What long, thin fingers he has—blackened from smoke and, apparently, cannonballs.

I take his hand and try to stand, but my knees are finding it difficult to cooperate. I crumple under my own weight, but not before I get a second look at that familiar face. Familiar, that is, if you're, say, looking through a book of art history.

"Cosa ci fai qui?"

"Um . . . how did I get here? Well, it's kind of a weird story. . . ."

His mouth curls in a secretive smile under his beard. If seeing is believing, I would have to confirm that the *Mona Lisa* has elements of a self-portrait after all.

He stakes the torch into the muddy ground. "The better to see you with, my dear." No, Leo didn't actually say those words, but I have to say, it feels like I'm in some kind of Grimm's fairy tale. Emphasis on *grim*.

"Where am I?" I hear myself mumble, although I'm thinking maybe this is a continuation of the dream I had in the Metro ride home from the Smithsonian.

"Dove sei?" asks the great da Vinci. *"A Firenze, naturalmente!"*

Are you kidding me! "Florence . . . as in, *Italia*? But a minute ago, I was—and then Lex was—but now I . . . and you!"

"Quale lingua parli?" This, again, from Leonardo.

"English. Um, er . . . I speak *Inglese*." Suddenly feeling panicked. I mean, I really don't know Italian, and the Latin . . . well, that's a dead language.

Hoping I can fire up the tablet and retrieve my Italian–English dictionary; luckily, my backpack seems to have cleared multiple time zones intact. But before I can unzip to check on the state of my electronics, Leonardo taps me on the clavicle.

My fuzzy headphones! They have survived the swirl of time and space!

Tuning in, I sense a repetition of an earlier message: *A tempo,* Carlotta. *A suo tempo.* Kairos's mindware! In its own time. But where is he? I can't see my hand in front of me, much less whether Kairos is within eyeshot.

In eyeshot, he echoes, and I breathe a sigh of relief. *And know this: Translator is an interpreter as well. While you wear this, you will not only un-*

derstand others, but you will be able to craft your reply in their tongue as well.

That would explain why Kairos speaks English so well. Except for that horrific accent.

"*Eh, bene.* Hey, *grazie* for the travel tips, Kairos!" I say aloud, grateful I had thought to demo the headphones before Lex's kiss sent me spiraling into deep time. The one eye that isn't swollen is finally beginning to adjust to the light-empty night, and the dense smoke from the cannon fire is clearing.

"Ah, so, you're English?" Leo's looking me over kind of strangely, like an alien has landed. And I guess I *am* an alien in his eyes! I look down and notice my ballet flats might seem a little out of place for his time. If it is, in fact, that time! My leggings are covered in mud, and my hair . . . well, let's say Gwendolyn Morton would not approve of the way her daughter looks in public today.

"No matter," Leonardo's saying. "What are you doing out so late, boy—you are but a boy, are you not?"

Wow, he doesn't realize I'm a girl! My wild mane wouldn't even signal my gender here, as Leo himself wears long, flowing locks. Introductions would be in order. After all, he might wonder if I'm hiding weapons or something.

"Pleased to meet you, Mr. da Vinci! I'm Charley . . . *eh, bene*, that is, Carlotta. Daughter of Gerald. Anyway, it's an honor!" I offer my hand.

"Carlotta! *Miracolo!* But you have interrupted my test, Carlotta. It is so rare to find a night without moon or cloud cover. The accuracy of cannon fire is *perfetta*, in the case of invasion by the French king, who even now plots against the duke. And my patron, Lorenzo de' Medici, will be angry with me. Another failure."

The earmuffs are burning my ears. I pull them off to cool down. Even without Translator, I think I'll get enough of the gist—at least for now.

"Sorry, *signore*." And I attempt a curtsy, as I imagine a girl from this time would be expected to show deference to a great master. But, clumsy Charley, I stumble over my own two feet, and once again, I am falling.

Leonardo extends a hand, and I find myself looking up into two luminous eyes. His expression has changed. *"E tu, bella signorina,* where did you come from?"

It occurs to me somewhere in the deep recesses of my befuddled brain that Leonardo is calling me *beautiful*. With his left hand he traces his long fingers along my cheek, and with the right, he lifts my chin outside the circle of torchlight, as if to find a bit of starlight to better see my face—like an artist might examine the bone structure of his model.

"And, speaking of sight—perchance we can calm the black and blue around your eye."

"Ouch!" I blink and jerk my head away, but not before I step hard on my bum foot. I wince.

"Swelling. I will have my apprentice fetch the appropriate ointments and herbs from the apothecary at first light. We must apply a soothing balm."

And it suddenly dawns on me: *Leonardo da Vinci is worried about me.* I mean, how cool is that!

This is so unreal. I feel like a player in one of Billy's virtual worlds, stumbling in the dark through smoke and cannon fire on a field where Leonardo himself is test-driving the future of night warfare. But it's no artificial intelligence responding to me. No prerecorded dialogue. It's him! Leonardo da Vinci, who was testing night weapons. Wonder if it would be an unfair advantage to tell him about the Stealth Bomber and drone warfare, or would that completely upend history?!

The Edge of Yesterday: My Coda

(Definition: a musical term that is Italian for the ending of a song or a dance—but also, like, The Last Word.)

And so, dear reader, my adventure begins. I have no idea what the future holds or, in fact, what future I can even look forward to. I have built a time machine spurred on by a drawing, a dream, and a school project. Even I couldn't imagine it actually working.

But then, out of the blue, I'm face-to-face with my Florentine idol, Leonardo da Vinci. I have no idea whether I will ever make it back in time to win the science fair, but I know I have accomplished something BIG. How big? Only time will tell.

It is only now, when Leonardo has disappeared into the inky night to retrieve his inventions—weapons for the Duke de' Medici's defense of Florence . . . missiles that I have designs for in his own drawings in my own time—that I have a moment to catch my breath. Luckily, the torch the Maestro used to illumine this spot a few moments ago has not yet burned out so I can scribble my thoughts here in my spiral for the science fair . . . or posterity . . . or something.

Whether you are reading this tale in the fifteenth or the twenty-first century, or somewhere in between, I can guarantee you, this is the adventure of a lifetime.

I am starting a private blog—top secret. Once I'm settled someplace

(I hope!) and *if* my tablet has survived the trip intact, I'll photograph my scribblings and upload them to the blog.

I am addressing it to Billy, because there is no one else who would believe what has happened. Not even my parents. ESPECIALLY not my parents!

If Billy gets it and decides to leak it—for my safety and your amazement—well, I cannot be held accountable for actions in a future I may have all but disappeared from!

Or maybe I won't even exist. . . .

> Blog Entry #1. FYEO
>
> *Space-Time Dimension (STD): @ today minus about 500 years, the Republic of Florence*
>
> *NEWS FLASH: IT WORKS! WE HAVE SUCCESSFUL-LY HACKED THE SPACE-TIME CONTINUUM. I know you won't believe me . . . I find it hard to believe myself! But I swear, Billy, somehow that little frame model you superglued, the golden ratio, and Kairos's algorithm all came together in one split moment to break the time barrier.*
>
> *I would send you a postcard to prove it, but Gutenberg's just barely in business and it is centuries before Louis Daguerre will invent the first real camera. Lucky for me, I have one on my phone! Meanwhile, I fear no one is aware of the outcome of the voyage of one New World explorer. In fact, I hardly know what's going on here myself; after all, it isn't like I can turn on CNN.*
>
> *Because I am sure no one but you could possibly believe the reality of this discovery, I've marked this FYEO—For Your Eyes Only.*
>
> *This is the first time I've had a chance to breathe. Literally.*

I mean, you try spiraling back in time half a millennium and let me know how you feel!

Anyway. Here goes our experiment in space-time blog transmissions. Luckily, I have my tablet. Must figure out if there's any weak signal of satellite transmission echoing in time to post this online and get to you.

By now, you may have heard that things with Lex in the garage didn't go so well. I have no idea how it happened, but first, I was showing him one thing, and next I know, I'm plopping down in an empty field in the black of night. Empty, that is, until this crazed guy with a mane as wild as mine shows up out of nowhere, covered with a hoodie and cloak, waving his arms like a madman and cursing his fortune—in Italian as it turns out. And you will never believe who the madman turned out to be!

Leonardo: the very guy I came to see. He asked what I, a mere girl, was doing out alone at midnight. A mere girl. And people say I have an attitude!

Of course, I have a million questions for the great, the magnificent, the Renaissance genius Leonardo da Vinci. . . .

Wow. Stay tuned, Billy, 'cause we're in the zone now!

Your Partner in Time Travel,

Charley

P.S. I am praying you receive this transmission and can ping me back.

If I can, I will send this as a blog later—that's a BIG IF. It seems a bit like putting a random message in a bottle in the middle of the ocean. Or random digits in the middle of space-time!

What was I thinking about when I came up with this idea—that I'd just bounce my signals up to a GPS satellite? I mean, Charley, dummy, there are no satellites in the fifteenth century. For all I know, Columbus still sails the ocean blue . . . at the distant edge of the known world. A lot like me!

So, if this post reaches Billy once I try to send it later today—I mean in *my* today, that is, my once and, hopefully, future Space-Time Dimension—what will he even make of it?

To tell the truth, I'm pretty scared. I mean, this whole thing is so much bigger than me. And my whole body's beginning to ache. I wonder if I might have some Advil floating in the bottom of my backpack, but I'm too tired to even look.

I begin to shake. The strange landscape suddenly blurs as my eyes well up, but I screw up my courage and set my mind—now that I'm here, I've got to complete my mission!

I feel so alone. Who ever could imagine our invention would actually work!

What am I supposed to do now?! I survived a major time shift, but will I survive this wild place and be able to learn from Leonardo—and even scarier, will I ever get home again?

Stay tuned, dear reader, 'cause the pit in my gut says the real test is just beginning.

BONUS CHAPTERS
BOOK TWO

I.

Discovering the Old World

The cannonball, the fall, l'Uomo Universale—it's like I've been in this situation before, but how can that be? There is nothing even remotely familiar in these surroundings.

I smell something like gunpowder, even though I've never really smelled gunpowder. What else could it be, this acrid stench? And after the cannonball rolled by my foot—and then I tried to move and felt my ankle start to throb . . .

This cannot be real, Charley.

I wipe the drip from my eyes. No tears! I look at my hands—grimy with soot and dirt. In my backpack, I find a crumpled tissue and use that to wipe off some of the schmutz. (This is a Yiddish word that means dirt or grime. It somehow sounds more descriptive when you say it out loud, *sch-mutz*. Try it!) Then I notice my tablet has the fingerprints of this dirty place all over the screen. I take a swipe at it with the now-dirty Kleenex and insert the tablet gently into my backpack. "Keep working . . . keep working . . . keep on working!" I beg the device. It suddenly feels like a lifeline to reality—whatever that is.

I take one last peek under the cover—the screensaver blinks reassuringly. Settled, I breathe a bit easier.

"*Al diavolo!*" bellows a loud, deep voice. I jump up as I hear heavy rumbling coming up from behind; my gimp foot strikes the ground, sending up shooting waves of pain, I suddenly see stars, and my whole body tenses —I can't take any more surprises.

"Ecco, Carlotta!"

"Huh?" So I wasn't dreaming. Here is Leonardo da Vinci, himself, cursing and calling out to me.

I squint and wink first one eye, then the other. One eye is still blurry, but at least I can open it a bit wider now.

Ser Leonardo (as I now know the one-and-only Leonardo da Vinci to be called) is approaching in my direction, tugging at a heavy rope, straining from the weight of a weird-looking iron cannon-like thingy mounted on a cart. I recognize it from a drawing in one of the da Vinci codices (plural for codex—meaning notebook), as one of his inventions: a triple-barreled cannon.

Weird. Who would believe that his designs for wildly future inventions might ever be constructed in Leonardo's day-and-age, much less actually work many centuries before their time?

But I am apparently proof that, whatever Ser Leonardo imagined, he could build—for now I have become history's first verifiable time traveler. His design for a time machine has brought me from the twenty-first century five hundred years back in time.

No mean feat for a fifteenth-century artist, dreamer, and inventor. Da Vinci, that is.

Now that I have achieved my own unlikely goal of meeting the world's original Renaissance genius, I have to figure out where I am, how I actually got here, and what to do while I'm here. And how to get back/wake up/get out of this reality show in time for the science fair. (Oh, did I mention: This all started out as part of a crazy, cockamamie scheme for the middle school science fair?)

Surveying the situation, I have to accomplish this feat with a maybe-broken ankle, a black eye, some random modern (for my time) high-tech gear that may or may not work here—wherever *here* actually turns out to be—and a handful of red, yellow, and green gummy worms to keep my tummy from rumbling.

Normally, I would use my senses—and my sense of reason—to figure out where I am and what the situation calls for. Unfortunately, panic, pain, hunger, and extreme disorientation have stripped me of my powers of reason. In biology, we learned that this triggers the primal "fight, freeze, or flight" response, like our ancient ancestors felt when a saber-toothed tiger was charging them.

Nonsense, Charley—you are not in danger of being mauled by wolves. (*But here is danger!* the voice inside my head shouts.) Just take some deep breaths. . . .

"We must get your ankle wrapped up, *cara.*"

Leonardo da Vinci's words first come to my ears as so much "blah, blah, blah," but, luckily, I have Translator—an invention that provides immediate translation from Italian to English, and vice versa—which lets me communicate here.

"*Cara,*" he said. I look around wondering who the Maestro would call "dear," until I realize he's talking to me! He carries a length of linen smudged with paint and smelling of pine trees. I wrinkle my nose to remember this scent—turpentine—and it comes to me: This must be an artist's rag he uses for painting. He motions to me to sit. "You will need both legs to walk into *Firenze.*"

"*Firenze, si!*" I say, peering through the darkness to register that this is, indeed, my current location. Something niggles at me about the idea of Florence—a feeling there is something else about this place. But my head hurts too much to call that to mind.

I try to concentrate on the here and now.

As he artfully wraps my foot in the smelly cloth, Ser Leonardo begins rapid-fire instructions telling me how to get to his *atelier*. (Translator helpfully indicates this is a French word meaning artist's studio or workshop.) "You must walk through this field, mount the stairs, and you'll find my studio at the top, over the tavern, *va bene!*"—all in Italian-accented Latin, or maybe Latinized Italian. Even with Translator and the rudiments (definition: basics) of a first-year Italian vocabulary, I am lost trying to follow directions. We all know, at best, I am "directionally challenged." And here, now, I am far from being at my best.

"*Adagio, molto adagio!*" I tell him over and over, then wonder if this is only a post-Renaissance musical notation for "slow down." I can't keep up when someone's talking really fast—like, apparently, Leonardo when he's excited. Like now.

Unexpectedly, I hear a horse's hooves pounding from a distance and someone shouting, "*Signorina!*" I turn around to a familiar face. Live and in person.

"Kairos!"

His horse circles around me once, twice, before he pulls in the reins to halt his steed.

"At your service, *bella* Carlotta!"

I want to hug him—I'm so relieved to see a familiar face—but the horse and Leonardo stand in my way. I hop up on my one good foot with what I'm sure is the silliest grin. "Kairos, I'm here! Me and the great da Vinci! And you!"

Leonardo looks from me to Kairos and back. "She is . . . ?" he begins, Kairos nodding in assent. "You have . . . ? *Puella ex machina est!*"

The girl from the machine.

"*Si, Maestro!* You did not miscalculate. The technology of the future is a marvel!" Kairos beams at me.

This adds to my confusion. "You're . . . ? So you meant me to . . . I mean you chose *me* to come here? How come?"

"Perchè no?" Leonardo exclaims. "You are curious. And smart. We also! Kindred spirits *ex tempora*. Out of time."

I think back to my first meeting with Kairos—during Take Your Child to Work Day at Dad's office last week (or 500-plus years from now, I remind myself). At first I thought he was someone's kid, there for the experience—he looked not that much older than me, although there are those who tell me I'm thirteen going on thirty. Then it seemed like he was some super-smart IT dude. And he admitted to being from another place. But this place?

"Show him, Carlotta! The magic slate you carry in your sack!" Kairos urges me. *"Maestro,* your wish is Carlotta's command. The world in a sack." Leonardo crowds closer, and I wonder whether I am not here to teach the master.

I unzip my backpack and start to sort through the junk inside: Solar battery, check. Lego bot servomotors, check. Gummy worms, check. Thank goodness I grabbed the backpack before Lex accosted me. (Guys and the games they play. Don't get me started!)

And therein, a bit of modern sustenance! I pull out the pack, extracting a single candy worm and suck in the familiar sugary sweetness.

A measure of calm returns, allowing me to think. What else is in here? I've forgotten. Cell phone, check. How the phone got in here, I can't remember for the life of me. I gently pull out my tablet, checking once again to make sure it's still working, and thank God, the Apple of Jobs appears. But this is weird: The time it shows no longer ticks by in exquisitely precise and uniform digits; an hourglass has replaced it.

When he notices the apple light up, Leonardo reaches out and touches the tablet reverently. *"Un dispositivo di conteggio?"*

"Um, well it can be used for counting. Among other things," I reply, thinking that to explain to the original Renaissance genius what a modern-day polymath (an all-in-one synonym for Renaissance genius) like Steve Jobs did in creating this device could take the next five hundred years.

"Hard landing?" Kairos asks, examining my bruises.

Before I can tell the story of my dramatic entry, Leonardo breaks in.

"How would she even know to find me here?" Leonardo asks. "Except for *il Duca*, all of Italy believes I am in *Milano*. After all, this is *Anno Domini millequattrocento novantadue*."

I don't get much of this besides the nickname for Lorenzo de' Medici—the Duke of Florence—and the city of Milan, and I'm guessing it's the year—hard to decipher, but I register it as 1492. Are we in Milan, then? I look around, squinting through the darkness to find a familiar landmark, but we appear to be in a countryside marked only by rolling hills, fields, and trees.

But the year . . . if it's 1492, what would be happening? I know Leonardo had commissions from the Duke of Milan . . . so that I would find him in Florence seems more than a coincidence. Kairos claps me a little too energetically on the shoulder; the push almost topples me over. I feel the sting of my foot as I try to regain my balance.

"A happy coincidence, *Maestro*," Kairos grins.

"That it is coincidence, I am not certain," Leonardo replies. "But of this, I am sure: Carlotta cannot remain here. What if she were to be seen like this, Kairos! *Il Duca* is already unhappy that I have missed his deadline for this confounded experiment in nocturnal warfare. I have barely been able to shoot my tri-barreled cannon to prove its superior firepower, much less improve its trajectory. I cannot disappoint further—not if I expect another commission from my patron."

I feel something nudging me next to my ear and find that Kairos's horse has detected the gummy worm source and is attempting to lick those remaining out of their cellophane bag.

As I attempt to pull the sugary treat out of his way, my gimp foot again hits the ground, sending a shooting pain up my leg. "Yowzers!"

"Shh!" Leonardo pulls my arm around his shoulder to steady me. "They may hear you! Kairos, you must find the apothecary at once; tell the man he's to meet Carlotta at my atelier. I have instructed her where to go."

As he sees my confusion, Kairos pipes in with a mindware explanation—a pause after each hurried pronouncement by the Master.

"He's trying to reassure you. He means to say, 'Do not fear, Carlotta. You are here to learn, are you not? And learn you shall, even as you may teach us much! But the way is not without danger. There are those who will condemn you knowing too much.'"

This does not seem to be an accurate translation. Leonardo is looking me up and down, looking for all the world like a disapproving father. He shakes his head. "Your dress is entirely unsuitable."

It's then I realize I'm still wearing the secondhand dress I thought would turn heads at the library—that ended up turning Lex's head quite a bit more than I imagined! I recognize my attempts at being fashion-forward seem a bit out of place now—even if I weren't mud-spattered. I can hear Beauty Queen Beth lecturing me on how tacky I look in this dress—in front of the painter of the *Mona Lisa*, no less.

Which begs the question of what *would* be appropriate. The only frame of reference I have is from the art of the Renaissance masters featuring curvy voluptuous nudes or sexy shepherdesses, or demurely draped madonnas (as in Mary, Jesus's mother, not the rockstar), or noblewomen in embroidered, close-fitting caps and pleated neo-Greek gowns.

But what, for a normal girl like me?

I am seized again with panic. What am I doing here, anyway?

Kairos must see my legs shaking as his horse clip-clops closer and begins to nuzzle me again, tickling my neck. He sticks his long, rough tongue out and manages this time to lick the gummy worm in my hand. I giggle despite myself and pull his nuzzle toward me, stroking it. It is comforting to know a horse is a horse, no matter what the century.

"You recognize my animal?" Kairos asks, using a warmer voice than I have yet heard from him.

I shake my head. I've only ever seen any of this as a computer hologram that I never quite thought of as real . . . except, I guess, for Kairos. Da Vinci was a guy from history. Though here I must add that Leonardo's supposed self-portraits do turn out to be quite a good likeness. But of course they would.

Kairos looks disappointed. "You don't recognize this snout"—here he pulls up his horse's nose—"and this?" He points to his own face and I mentally flash on the *modello*.

"It's . . . you're . . . he's?"

I root through my backpack once more and my hands close in on a newspaper-wrapped miniature sculpture. Yes, it's in here! I carefully draw out what then seemed a novelty, given to me in another time by one self-proclaimed alien, a.k.a. Kairos. Unwrapping the newspaper slowly, I note thankfully that the sculpture has survived this wild ride intact.

Kairos nods. "Carlotta, fear not. You will be safe as long as you follow the *Maestro*'s commands."

I feel like a weight's been lifted off me. Kairos, the formula, the golden compass, the maquette. I would think it was destiny—*il destino*—if I believed in the irrational, illogical evolution of a universe in which time may flow in any direction. Which I don't, of course. At least, I think I don't.

But Leonardo, on seeing his mini opus, grows even more impatient. "Show no one these treasures of yours, Carlotta. You will be in grave danger! If they were to find you here . . . it will be light before long, and we must keep your secrets safe.

"So go—now, *sbrigati*! You will find the wife of the tavern keeper below my studio. Signora Vincenzo, by name. She has four daughters and servant girls of various sizes and shapes. She will be able to find you suitable clothing. And you have not eaten?"

I'm embarrassed to think he must hear my tummy growling. After all, I have not traveled five hundred years across six time zones without feeling a pit in the stomach. Gummy worms are not gonna cut it. *"Ah, si, Ser Leonardo! Buon appetito!"*

"Va bene, Kairos!" Leonardo commands. "Fetch the apothecary at once. Do not share with our good doctor any of the true details of your mission, lest he suspect strange occurrences. Know—he will not be happy to be wakened from his slumbers, but you must get her salves and dressings so there will be no swelling to the brain."

The brain! I touch my head gently, feeling the start of a good bump on the forehead. Yowee!

I start to say let's pick up a hot-and-cold pack and ibuprofen at the CVS—but then I remember where I am.

Kairos circles his trusty steed close again and taps me on the head, reminding me of Translator. "Remember to listen carefully to all voices around you. You will be able to tune in to all you need to hear," he whispers.

Before I can respond, Leo slaps Kairos's horse on the rump to send my friend out of time and into the night.

"And you!" I quickly come to my senses as Leonardo turns back to me, waves those artist's hands with those long, amazing fingers, and warns, *"La vedova Vincenzo* will be tending the fires soon: You must hasten there before dawn to avoid suspicion."

"But what do I say if she asks where I've come from?"

"Tell her you are Kairos's English cousin. You speak only the language of the British Isles. And do not share any of your strange intelligence with others. If they learn of your powers, those religious men of great power will stop at nothing to destroy you—and your knowledge."

"The powers that be would destroy . . . me? But what threat am I to them?"

"You see the future. And it is unlikely to please those who want to control our knowledge of how things really work! So be stealthy with your strange tablet," Leonardo warns, as he removes his cape to wrap it around my shoulders.

Smoothing my hair from my face with fingers that seem to know every bone and hollow, he throws the hood over my head. It's way too big; I feel like the disappearing girl, lost in its folds.

Blinded momentarily, I push back the hood as my fingers feel for a tie or buckle to fasten it more securely. Just then, I feel a tap on my behind. I turn to see Leo pointing, and I hobble off in darkness on a path that will lead I know not where.

II.
Do You Take American Express?

The distance is longer than I judged, especially with my bum foot, no moonlight, and certainly no electric lights. There are no people here. I pull Translator down around my neck—the better to hear.

Although my eyes have adjusted somewhat to the night, the swelling around my eye makes it harder to see where I'm going. I stumble over every rock and tree root, landing on my bottom more than once. It's discouraging. It takes a moment to regain enough energy to move on. I cannot help but look up at the vastness of the sky and wonder how I got here. Though I know I must hurry, in a way, I'm glad: Who ever saw so many stars!

I cannot stop to stargaze now; I'm in danger—or so Leonardo informs me. For a girl used to suburban whirs like air conditioners and cars, these strange night sounds—a hooting owl, the moo of a cow, a wolf howling in the distance—conspire to keep me moving.

I fit Translator safely back over my ears to muffle these inhuman sounds. Besides, its muffs are warm and there's a chill breeze blowing into the folds of the cape.

I spy the stars glittering off the Arno River that I see is not more than ten yards from me and lick my parched lips. I wish I'd thought to bring a filtered water bottle. Who knows whether it's safe to drink from the river!

As I approach the bridge, I spy two men idling by the road, one holding the lead of a mule. My brain registers the order, *Hide!* But I'm too tired, and besides, I am lost. Maybe they can confirm I'm going in the right direction. Or at least they might have a canteen of fresh water I can sip.

"*Scusi?*" I say tentatively.

The strangers apparently are waiting for such a moment. The first guy limps over with a smile.

"*Eh, signorina.* What is one *si bella* doing out at this tender hour?"

I curtsy, as I did with Leonardo.

"*Scusi, signor. Dov'è la casa di Vincenzo, vi prego?*"

I am not too tired to notice the look he exchanges with his *compadre.* "Aha! Antonio, it is Vincenzo she is looking for! Shall we lead her there?"

Antonio comes closer, leading his mule. I wish I could throw myself on that animal and ride. My feet are hurting! I knew I should've worn my Nikes instead of these stupid ballet flats. "*Oh, no! Sneakers with that outfit?*" I can imagine Bethy lecturing me. I feel a pang. Where's Bethy now?

"*Eh, perché no?* Vincenzo is on our way, no?" He grabs me by the arm with a leering grin.

Fatigue or no, I swat at his face, screaming as loud as I can: "Let go of me! *STULTUS ASINUS! Stupido!*"

"This one is lively," Antonio grins, ducking my slap. His friend comes over and removes Antonio's hand from my arm.

"*Eh, madonna!* We would not hurt you! *Amici*—we are friends. *Vieni, madonna!* You are safe with us. Here, would you like to ride this gentle animal? It is unseemly to make a *bella donna* walk such a distance. Allow me. . . ."

The first guy cups his hands and motions me to step up.

I hesitate: These guys look smarmy. But no stranger than anything else I've encountered in this unlikely voyage. And I'm sooo tired. "Well . . . it is kinda farther than I thought. . . ."

I step into the cup of his hands with my good foot, shoving the length of cape out of my way. When I realize using this foot would put me onto the mule facing backward, I pause to test whether my gimp ankle will support my weight—then out of the corner of my eye, I see a glint of metal flashing in the starlight: Antonio has a knife. He grabs Leo's cape—and my backpack along with it.

Cutpurses and thieves! With a sudden rush of adrenaline, I swerve and kick the knife out of his hand. He yells, cursing at me loudly, while his friend stops to pick up the knife. I only have a second to save myself. Grabbing my pack, I make a dash for dear life, mindless of my ankle. Even as I'm running, I manage to strap the backpack to my stomach and refasten Leo's cape around my shoulders.

I hear a donkey bray and one of the men panting in close pursuit. I have to lose them! I head for a grove of trees on the bank of the river, with no idea where I'm headed, except away from Antonio and his evil friend. At that moment, I notice a band of what must be gypsy families walking or pulling carts, carrying torches, and headed caravan-style across the bridge.

Not seeing an alternative, I scream at the top of my lungs, "HELP! RAPE!!! *Carnale!*"

A woman shrieks and grabs her child close for safety. Two men, colorfully dressed and carrying sacks on long poles, peel off from the gypsy band to charge at the cutpurses with their sticks.

Like a cloud of locusts, the women and children begin swarming around me, closing ranks as the men run in pursuit of the *banditos*. Their nearness, carrying along with them strange noises and smells, makes it difficult for me to breathe.

"Stop!" I shout, kicking, weaving, and ducking beneath arms attempting to grab and snatch at me, my hair, my cape.

I wrap the cloak tighter, looking for a way to escape the crush of the crowd, and run as fast as I can across the wooden bridge to continue on the path that runs parallel to the Arno on the other side. I pray it's the one that will take me into town. Now the path is narrow, and stepping off of it might mean slipping down the steep banks into the cold, coursing river.

I slow my steps to avoid disaster. After what seems like an eternity—in the way that finding your way in a strange place without landmarks or, let's face it, GPS, always seems never-ending—I begin to see signs of life.

A quick look around: The veil of night is brightening and I begin to gain my bearings. This, finally, starts to look like the iconic postcard images of Florence I remember from home. Funny that I've wanted to visit here my whole life and, as luck would have it, by the time I make it to Florence it's in a whole 'nother century. While I am happy to be here, I wonder what it would be like to visit in my century, with my parents. And come to think of Gwen and Jerry, are they even aware that I've disappeared?

Homesickness fills my heart as I remember my mother is also in Italy, playing violin on a concert tour with the National Symphony Orchestra—almost near, and, at the same time, oh, so far away!

I swallow past the lump in my throat. I stop to catch my breath, then slide to the ground, carefully stretching out the smelly, rag-bandaged leg to protect my bad foot.

The picture-perfect landscape of Florence blurs beneath my tears. Feeling sorry for myself is not going to help anything; meeting Leonardo is a dream come true. I am here to find out the secrets of his incredible mastery of art, engineering, science, invention—in short, everything under the sun. And if there's anything Gwendolyn Morton taught me it's this: What I can dream, I can do.

I wipe my eyes with Leonardo's cloak, overly large and falling off my shoulders. I move forward, inspired.

Finally, I come upon a building with a large doorway, a public house. I enter the courtyard. There is a simple wooden sign over the tavern door in Italian: PER CHI DESIDERA TRATTENERSI E ESSERE ALLEGRI. I know from music that *allegri* means joyful, so I can't imagine it says anything bad. I also notice steps and a ramp that lead to the upper level, undoubtedly Leonardo's studio. I pound on the door, yelling breathlessly, "Someone! Anyone? Is Vincenzo there? Kairos sent me! I am his American—er, English—*cugina.*"

A set of eyes peers out from behind a peephole in the door, grunting something incomprehensible. It's then I realize I am no longer wearing Translator. Did it fall off when I was fleeing those awful bandits? I can't have lost my only means of communication!

The woman hisses, *"Shh, regazza! Vi sveglierete le protezioni de' Medici!"*

De' Medici. It's the only word I can grab on to. The Duke of Florence is Leonardo's patron, a man of enlightenment. Yet this woman seems afraid.

"I need to find Signora Vincenzo!" I sob. "Please. I mean *grazie!* Um, err, *prego?*"

I feel my lips begin to quiver and a sob escapes. At that, the woman cracks the door a tad wider, curious to get a closer look; flickering candlelight throws wild shadows against the wall.

"Leonardo *dice . . .*"

As if I've uttered a magic word, I hear the jingle of skeleton keys followed by the click of door bolts. A wizened red face framed by a white cap and wisps of graying hair peeks out guardedly.

"Io sono la Signora Vincenzo," she says.

"Oh, thank God!" I say, and grab her hand still holding the keys. Still, she will only open the door enough to look me up and down.

I brush off the hood with my free hand and quick pull my hair out of its ponytail so it falls around my shoulders. It's then I feel something weighing down my hood. Translator! I put on the headphones and fluff my curls around them to mask the purple muffs, something that would undoubtedly raise suspicion. For once, I am happy to have big hair.

"Why, you're a tiny thing, aren't you? What are you doing out *di prima mattina?*"

I breathe a sigh of relief: The translation into English is coming through. I pray that Translator is working in both directions.

"Those thieves—they tried to steal my backpack and I ran, but they had a donkey and they're probably coming for me. And I'm starving! Maestro Leonardo said—" Feeling desperate, I grab her hand again. "Oh, please do let me in!"

As she pulls her hand from my grip, she says, "Oh, come now. This is not the face of *una brava. Come ti chiami?*"

"Charley. I mean, Carlotta. As soon as Kairos gets here—"

Uttering Kairos's name earns a look of disapproval.

"Carlotta. You're a wild-looking thing. Kairos, you say? That boy is always finding wayward wenches. No matter, his heart is in the right place. Now, it is three florins a night for a bed." She holds her hand out with a frown, stubbornly blocking the door. "Payment in advance."

"Pay? A bed? Geez, I mean . . . I'm not going to sleep here or anything. But I hadn't thought . . . where . . . ?"

I begin weeding through the front pocket of my backpack for spare change, praying to heaven that something resembling currency might be tucked away. I spent most of my last allowance on supplies. I dig up a few pennies and my lucky Susan B. Anthony dollar. In the process, out spills random colored Post-it notes, paper clips, energy bar wrappers, aluminum foil, and ballpoint pens. A colorful braid of gummy worms, moistened by horse slobber so they are now fused into one solid, squiggly snake, falls to the sooty ground at my feet. Ruefully, I think, there goes my quick comfort food.

"No, no, NO!" I am startled at her sudden screams and look up to see a terrified look cross Signora Vincenzo's face. "Who are you anyway, young daughter of Eve, carrying snakes that fall from your sack? *Strega!*" She kicks up the dust as if to bury the snake but then realizes it is not alive.

I bend to pick it up, dusting off the sticky thing and pretending to chomp down. "No, it's just candy. Or was. You know, sweets?" I say, frowning. Seeing the inert sugary mess in my hands, she leans in closer, then frowns.

"*Si tratta di uno scherzo?*"

"No, not a joke . . . and unfortunately, I don't have any euros. I should've asked my mom before she left for Flor—I mean, before I left home. Would you take a Susie B.?" I reveal the bright silver coin in my palm. "Or maybe you take credit? We get cash back. . . ." I flash my American Express card—per Jerry Morton, for emergencies only, and if this isn't an emergency, I don't know what is—before realizing, once again, there is no America, no credit card industry, no Dad. Not here.

"No, *madonna*, I do not give credit to strangers," she sniffs, arms folded across her chest. "We are cheated by too many of the likes of you!"

"I'm good for it. I swear! Please, take the money. It's all I have right now, but I promise, I'll figure out how to get you more." I have no idea how this might actually happen, but I need to go inside and sit down. Get off my sore foot, at the least.

Signora Vincenzo ponders this momentarily, then takes the Susie B., holding it up to the light. Placing the coin between her teeth, she bites.

So many germs, I want to warn her. Doesn't she realize where that's been? But no, of course she doesn't. I see a greedy look dawning on her face.

"*Si. Viene,* Carlotta. You are English?"

"*Si, o no.* Well, it's sort of *difficile da* explain. I was in the garage showing Lexy the thingy, and he tried to kiss me. . . ." I pucker my lips. "Ugh! Then somehow, I must've bumped up against something that switched the machine on, but it was an accident, and suddenly, I was spinning . . . and Lex was gone!" I can't stop spewing.

"But then here I was . . . er, am . . . and there *he* was. Leonardo, that is. Firing cannonballs! He said I had to come here or Lorenzo would arrest me, but I need to talk to Leonardo some more and I can't go to prison!"

Signora Vincenzo looks perplexed. She lets out a long sigh of frustration.

"*In veritas,*" I try. "I won't rob you or anything," I promise, holding up three fingers in the Girl Scout pledge. I hope she recognizes "I swear" as some sort of universal signal. "And Leonardo is also expecting me." I feel my lip start to quiver again.

"And really, my foot—I twisted it. And this bruise . . ." I pull up the cape to show her. My ankle's blue, and swollen to baseball-size now. "Ser Leonardo sent Kairos to get something to calm the swelling. Could I sit down?" I pantomime sitting, to make sure she *capisce.*

This whole conversation isn't going well. It seems we are lost in translation, technology or no. I put my hands to my ears just in case Translator has fallen off again, and am only slightly reassured to feel the fur.

"You know, inside, maybe?" I'm starting to get a headache now, too. Whether from the bump on the noggin or the long trip, I can't tell for sure. But I do know that if I have to stand up much longer, I might faint.

"Enter, *signorina*," she commands after a long pause. "Then you aren't one of those *pazzi* girls? They're chasing Kairos day and night. They always land at *mio porta*."

One of those crazy girls. Kairos is as bad as Lex, it would seem.